A People's Green New Deal

"Hands-down the best book yet on the Green New Deal. Courageous, bold, refreshing – Ajl envisions an ecosocialist transition that is rooted in principles of global justice."

—Jason Hickel, author of *Less is More*

"Ajl guides us with an authority steeped in scholarship but also with panache. If you really want to learn what'll be necessary for our species to survive climate apocalypse, read this book. You'll then know the ways by which humanity's very fate can be won."

—Rob Wallace, author of *Dead Epidemiologists: On the Origins of COVID-19*

"Anyone wanting to understand the limitations of the Green New Deal, and how it is being employed as a tool to rationalize Green Capitalism, and sanitize its advance within the capitalist system must read this critical work."

—Kali Akuno, Executive Director of Cooperation Jackson

"You cannot purchase your way out of climate change the same way you cannot pick a 'Green New Deal' brand that suits your personal preferences, Ajl argues. Neither are real alternatives. Anti-imperialism and anti-capitalism are not bygone projects. They're very much alive in the South and at the center of climate movements. Left climate movements in the North would be better served by following their example as well as reading this critical work."

—Nick Estes, author of *Our History Is the Future: Standing Rock Versus the Dakota Access Pipeline, and the Long Tradition of Indigenous Resistance*

A People's Green New Deal

Max Ajl

PLUTO PRESS

First published 2021 by Pluto Press
345 Archway Road, London N6 5AA

www.plutobooks.com

British Library Cataloguing in Publication Data
A catalogue record for this book is available from the British Library

ISBN 978 0 7453 4174 3 Hardback
ISBN 978 0 7453 4175 0 Paperback
ISBN 978 1 78680 705 2 PDF
ISBN 978 1 78680 706 9 EPUB
ISBN 978 1 78680 707 6 Kindle

This book is printed on paper suitable for recycling and made from fully managed and sustained forest sources. Logging, pulping and manufacturing processes are expected to conform to the environmental standards of the country of origin.

Typeset by Stanford DTP Services, Northampton, England

Simultaneously printed in the United Kingdom and United States of America

Contents

Acknowledgments

Thanks to my editor, David Shulman, for supporting this project from the outset. Thanks to many comrades, some of who did not know me very well, for offering very quick feedback on my draft manuscript or suggestions for sources: James Trafford, Nemanja Lukić, Stefania Barca, Bue Rübner Hansen, Cora Roelofs, Matt Haugen, Alex Heffron, Willliam Thomson, Brant Roberts, Eric Holt-Giménez, Sam Markwell, Zeyad El Nabolsy, Fadil Aliriza, Mabrouka M'Barek, and Justin Podur. Thanks to Anitha Stephen for preparing the index. Thanks also to Rob Wallace, Maywa Montenegro de Wit, Mili Narayen, Aaron Vastinjan, Ruth Nyambura, Rut Elliot Blomqvist, and John Gulick for ongoing discussion on eco-socialism.

often landless, displaced to slums, forced to work in factories that produce what Amazon sells.

A Green New Deal (GND) which imagines replacing delivery vehicles which run on oil with those that run on lithium batteries, which imagines a continued closed-loop consumer economy based on recycling every bit of metal and plastic and replacing natural gas with state-subsidized nuclear power, could work for Amazon's business model and the capitalism in which it fits. It would be no problem at all for Bezos – at least in the short term.

But a People's Green New Deal based on universal access to renewable energy, climate debt payments, de-commodified public spaces, funding for the arts, 24-hour public libraries, and food sovereignty in the South and the North alike would wreak havoc on Amazon's supply chain. It would empower the laborers who fill its lower rungs. It would chip away at the coerced and individualized and debt-fueled consumption on which it rests. And it would sever huge portions of people's needs from capitalist accumulation circuits. Moving to agroecology, or "the application of ecological principles and concepts for the design and management of sustainable agro-ecosystems," would further break apart the Amazon-friendly structuring of the world.[2] If the People's Green New Deal enfolded demands for political sovereignty in the South, such a plan for planetary just transition would burst apart a capitalist system which rests on super-exploitation of the Third World, using it as a garbage dump and labor supplier. A People's GND would become eco-socialist: the worldwide decommodification of social life and the conscious and conscientious management by the associated producers, or revolutionary humanity, of our relationship with non-human nature so as to ensure its longevity and well-being.

As Bezos says: it's probably unhelpful to say too much about the GND in general. Angels and devils dwell in the details. Which GND and why does it matter? What is at stake in one or the next, who has a stake in one or the next, and for whom does it not matter at all? This book is primarily about the GND debate in the United States. But the US resembles an island only in the most literal sense, and the same debates are happening across the imperial states. More importantly, everything the US political elite does or does not do occurs and impacts a sharply uneven world – from Pentagon budgets and the petrodollar which translate into a fire from the sky on Yemeni smallholders, to the decision to demilitarize, to speeds-of-emissions reductions which submerge or stabilize the Seychelles. And from cobalt-laden tech for batteries for fleets of electric cars which rely on the continuing underdevelopment and war-fueled annihilation of the Congo to keep their prices cheap, to the decision to quickly move the population of the US to free electrified public transport, bicycles, and local work.

Any GND aims to change physical production, social power, and how humans interact with the environment. GNDs are plans for preserving, strengthening, remolding, attacking, or abolishing capitalism. None of the GNDs, or programs for degrowth, or programs for green capitalism, restrict themselves to just fixing up a mutilated natural world, or imagines a simple swap of lithium batteries for roaring internal combustion motors or windmills for coal-fueled power plants. GNDs are agendas for governing, for deciding who gets what and how much, who does not get, or how little they get. Since the US economy is connected through its monopolies and its military to oppressed nations and classes all over the world, a US GND is an agenda for governing the nations of the planet. It could also be something else: not an ecological empire but empire's end, the planet's other nations ruling themselves, including colonized nations struggling against settler-colonialism.

By now the term Green New Deal spans a sweep of proposals, from the European Union's GND for Europe, to the dead-letter Bernie Sanders GND, and on to the degrowth-oriented European GND emanating from the nursery of cutting-edge political ecology, Lisbon and Barcelona, also the urban cores of Europe's own Third World. Then there are the eco-socialist plans of Howie Hawkins and the US Green Party, and the Red Deal for indigenous decolonization from the US's Red Nation. Some imagine a thorium-and-solar-fueled US military. Some imagine no US military. Some plan for a revamped US workshop-of-the-world exporting clean tech to the Third World. Some fight for climate reparations steadily erasing the economic apartheid between First and Third Worlds, making in its place a world with room for many worlds.

Green New Deals which seek to help other nations shift to cleaner forms of production are GNDs which are more democratic, humane, and just than those which do not, and those which seek to profit from such "help" may not change very much at all. GNDs having a worldwide shift to agro-ecological forms of farming based on food sovereignty and far less trade as their touchstone, and which argue for local democratic economies built on appropriate technologies and sovereign industrialization and local control of renewable energy, catering to human needs, are transformative. They are not compatible with capitalism.

This book clarifies what is at stake in different Green New Deals, whose politics they accommodate and reflect, and how big capitalists might work with or against distinct GNDs. Most fundamental to each GND in the Euro-American sphere, is its relationship to capitalism and imperialism: Are these agendas for governing capitalism, or for destroying it? If the answer is the former, pragmatism and realism can become firehoses dousing in despair the flames of revolutionary hope. If the answer is the latter, as it should be, then questions of "realism" should be dismissed as ideological counterinsurgency against those who not merely hope for, but need, a better world.

WHY ARE WE TALKING ENVIRONMENTAL POLITICS NOW?

Why now? After all, the Green New Deal is not a new idea. It is an idea that has been born, reborn, and born again, aborted and living half-lives for 15 years. Although it was bandied around in various circles for many years, it first emerged in the public sphere in that petri dish for Eurocentric bacteria, Thomas Friedman's *New York Times* column, wherein Friedman paused from calling for blitzkriegs against Iraqi children to call for a Green New Deal. Friedman was frank. It was meant to be "geostrategic, geoeconomic, capitalistic and patriotic," because "living, working, designing, manufacturing and projecting America in a green way can be the basis of a new unifying political movement for the 21st century."[3] This was GND as meta-frame for industrial renaissance, shoring up weak points in the edifice of US power. Friedman as the in-house priest of US imperialist neoliberalism was not then or ever shy about telling his readers what he thought should happen or who would benefit. Nevertheless, 2007 was premature. 2008 saw President Barack Obama stumping with mention of a GND for the United States and diverting some of the federal recovery budget to alternative power sources.

2009 saw Europe push a GND as an agenda for rule, development, foreign affairs, domestic reweaving of a frayed social safety net, and other politics that sought to impale neoliberalism and mounting CO_2 emissions with a two-pronged fork: new spending for the people and new investments to rapidly ramp down CO_2 emissions in favor of new and clean sources of energy. It called for "economic, social and environmental solidarity," and a Europe acting "for its citizens" in lieu of "narrow industry interests," alongside a "new economy driven by long-term prosperity, not short-term profiteering," decent minimum living standards, small increases in development aid, and massive R&D. Finally, "Social and sustainable development clauses in trade partnerships should therefore be binding."[4] This was an early taste of the tendency towards apolitical social management – the governance approach – which tries to dissolve contradictions around class and colonialism, and capitalism's inability as a historical system to respect the earth-system and the tenuous, delicate, and easily shattered niche it has for many billions of humans. Long-term planning and the equally vague call for "decent minimum living standards" all gesture towards using the GND as a lever to shift the social system to "sustainability," another voguish word adulterated to meaninglessness through technocratic, bureaucratic, and academic misuse and overuse. The formerly colonized world and its rights to reparations, restitution, and development, summarized in the short and brilliant keyword of climate debt, were not present in this early precursor of current social democratic GNDs. Rather, they imagined managing South–North inequalities not by any kind of recognition that the empire of free trade produced the world-system, but by whittling away at structurally lopsided exchange through this or that melioristic regulatory clause.

The post-2008 recovery, partially based upon small state investment in the green industrial sector, silenced those amongst the vanguard capitalist class agitating for a new system. Neither egalitarian nor ecological, but a new regime with new frontiers of accumulation. Bubbles, quantitative easing, and low-level Keynesian spending by the US and China drowned out anti-systemic talk at the surface level of politics.

But 2008 also produced a kind of quiet revolution. It did not collapse the structures of capitalism nor make good on the arrears of colonial debt. But it jolted the work-a-day hegemony of capitalism. Just ten years after Seattle had delegitimized the World Trade Organization, the 2008 Wall Street Crisis led many to question whether capitalism was indeed the end of history as imagined by its apologists. In 2011 the occupation of Zuccotti Park and the US-wide Occupy movement again questioned and actively challenged capitalism as a political method of rule. Those who lived through these moments and saw the plate-glass shop window of capitalism's fixity shatter in front of us began and continue to organize.

THE POLITICAL MOMENT

The post-2011 period has been often characterized as one of leftist resurgence. The catastrophic worldwide advance of imperialism during this period has shattered entire states in the Arab world and emplaced conservatives through coup after coup in place of the Latin American radical governments which had been the political fortresses, partially conquered by the people, within which the Cochabamba meetings could assemble or the peasant movement La Via Campesina could grow stronger. These facts should make us hesitate to accept the narrative of radical renaissance. A more correct description would be a slow attempt to restore the level of left practice, theory, and organization achieved in the late 1990s and 2000s, including, with stalactite-slowness, some kind of understanding that people in the North should oppose imperialism. This has manifested alongside far wider unease, taking organizational form, for example, in the Democratic Socialists of America, with capitalism as a system of rule and accumulation, and far wider acceptance of some sort of socialist alternative.

An element of that older movement-of-movements was a rupture with regulatory-technocratic approaches in the North to dealing with climate change. Up until that point it had been the object of neutral policy chatter of emissions controls and carbon credits. From 1999 onwards, North and South started to surge towards one another. Increasingly, climate debt was on the tip of every activist's tongue, not as a slogan, but as a common-sense way to begin to repay the historical value torn from the South. While *alter-globalization*, or the idea of a different people's globalization, at some level knew what it was against more than what it was for, whether ideas of worlds within which many worlds fit, or Samir Amin's unity in diversity, it did not congeal into a political movement. The disintegrating and moderated forces of foundation funding, the political counterinsurgency of Democratic Party elite capture, and the rising atmosphere of judicial persecution against those who raised the anti-imperialist pennant or dared even to channel funds to children in the Gaza Strip, carried out a leveling operation in the post-9/11 atmosphere of rising domestic repression.[5] 2016 and onwards has seen, for better and worse, a moment in which crises of organizing and theory coincide with the rise and defeat of great yet greatly flawed and ultimately defeated social democratic avatars Jeremy Corbyn, Jean-Luc Mélenchon, and Bernie Sanders.

This rise-and-fall has also occurred as huge swathes of even domestic populations in the capitalist core are prepared for a more radical, transformative message – one that includes building with rather than ignoring the needs of the social movements and left governments of the South, and which accepts that we in the North also have a unique burden in carrying through worldwide social transformation, specifically in preventing our own governments from imposing by violence their political values on other countries.

These forces have emerged simultaneous to a climate crisis that has broken through into public view. As a result, many forces are seizing on the climate-development-social welfare nexus as a way of packaging their agendas. Against this background, newly elected Congressperson Alexandria Ocasio-Cortez ("AOC") joined with Senator Ed Markey to introduce a non-binding resolution calling for a Green New Deal. With quite a bit of marketing, a millennial fluency in social media, and the lurking institutional backing of the Justice Dems project, an effort, linked to the Sanders candidacy, to put progressive candidates into office, and which plucked out Ocasio-Cortez and crafted her candidacy, she introduced the GND, a concise nearly bullet-point piece of non-binding legislation. I will have more to say later, but for now, what is important to keep in mind are four things. One, Markey/AOC welded climate crisis to social crisis: the 40 years in the desert of "wage stagnation, deindustrialization, and anti-labor policies" appeared alongside the need to keep the climate below 1.5°C of warming.[6] Greening social policy or painting environmentalism red is not new, but here became part of setting the stage for a transformative policy package. And grounding social policy in emergency ecological crisis adds urgency, even makes for an emergency. Two, the legislation calls for a "new national, social, industrial, and economic mobilization," along the lines of World War II. Three, the legislation is not socialist. It does not claim that the people or the working class are in antagonism with capital. It instead calls for "transparent and inclusive consultation…and partnership with…businesses," and calls for directing "adequate capital… [to] businesses working on the Green New Deal mobilization," alongside a murky call for public "appropriate ownership stakes and returns on investment" in such capital grants. And four, while it speaks of the "historic oppression" of the poor and low-income workers, it calls for resolving that oppression while keeping the fundamental property structures of the country intact.[7] What has been remarkable, then, is how much people have made of what little was said, and furthermore how little attention has been paid to what the legislation actually told us. Capital grants to businesses is exactly the existing policy of US capitalism. To green it is to call for the greenwashing of capitalism. To enfold those policies in warm comfort blankets like "frontline communities" may hide the aims for accumulation at their core but does not change what is inside.

POLITICIZING CLIMATE POLITICS

This book does not defend the AOC/Markey Green New Deal, nor the forces behind it, which need to be addressed forthrightly. Whether that GND has advanced the conversation is a different question, and I am not very sure: at the moment, the forces in the driver's seat of environmental policy are green capitalist/post-capitalist transition agendas, and many who wish to "push left" the AOC/Markey GND may well end up disorganizing resistance to the broader ruling-class agenda by embracing a false opposition to it. On the other hand, the Red Nation's Red Deal and the Democratic Socialists of America Green New Deal Principles, each anti-imperialist, anti-capitalist, and in favor of climate debt settlement, stake out far more radical positions.[8] I write here then primarily about what the AOC/Markey GND explicitly states, and affirm that a more radical GND is possible. But possibility only emerges when we have greater clarity about what else is on the table and whether those forces can be allied to us – "us" being people who believe in certain universals: everyone in the world deserves free and more or less equal access to electricity, health, education, culturally appropriate food, housing, and reasonable and un-alienated work.

The existing debate about the Green New Deal and climate change tends to follow one of several depoliticizing, opportunist, or reformist analytical approaches. In saying reformist, I do not mean to stake out a position against reforms. Decommodification of production and consumption, or making it so that things we make can be made outside of the schedule of priorities imposed by what "the market will value," and the things we need we get automatically by virtue of living in a community rather than something we must have money for, is a good goal, wherever and whenever possible. And making sure that doing so does not damage the non-human environment is a good goal, whenever possible. Clean technology grants, including know-how and turning intellectual property rights into the commons of humanity as one form of climate reparations are good things. Reforms are not the same as reformism.

I mean by "reformism" analyses and political programs which refuse to consider that the world we live in and which we analyze and which we want to take part in liberating exists within an ongoing and world-spanning history of class struggle occurring in a divided world.[9] An analysis that refuses to consider that the Sana'a or Sheikh Jarrah slum dweller has an interest in ensuring that any "Great Transition" ensures their own liberation, and does not account for how accumulation on a world scale rests on the wreckage of Yemen and the wealth wrested from Sri Lankan tropical monocrops, is flawed. It leads to reformist politics by aiming only for compromise. In this sense, the somber hard-headed pragmatic notion of realism quietly states that certain kinds of suffering and oppression ought not to be dealt with for now. These are all symptoms of opportunism: to politicize climate questions, but only so far. These plans are not eco-socialist.

The common thread of opportunist, reformist, and social democratic approaches is a blind faith in technology, a kind of magical catalyst, stardust sprinkled on the current system and capable of transforming it into a just and sustainable world ecology. I examine this in much more depth and length in Chapter 2, where I discuss green modernization as the fundamental mythology and political practice of domination in an age of climate crisis. Green modernization theory, the touchstone of much of the Great Transitions' literature, the subject of Chapter 1, which covers plans to rework accumulation so as not to wreck the climate, accomplishes a similar project for an age when technology and the ideology of technology must be rethought and more to the point, reimposed, as a component of the inevitable march of progress towards a green horizon. Modernization theory, including and especially green modernization theory, the subject of Chapter 2, rests on a foundation made of quicksand: the myth of progress, the idea that things overall have gotten better and will continue to get better. This is perhaps most apparent in ideas for various geoengineering gewgaws, which stand in for immediate, sharp, socialist transformations in core living patterns alongside immediate payments of ecological debt. I reject the approach of urging the poor to pay the cost of empire as a way of life, a point of convergence between non-revolutionary GNDs.

A PLANETARY GREEN NEW DEAL?

In thinking through a planetary GND which can accommodate the demands of the majority of the Earth, we should start with earlier proposals that came from the majority world and were touchstones for climate activists before "GND" became a household term and a common shorthand for climate-change politics. This takes us back over a decade, to when the Cochabamba people's process took place, setting the framework for climate debt discussions. The process emerged from the 2009 success of the peoples of the South. Radicalized states like Cuba, Venezuela, and Bolivia blocked the

United Nations Framework Convention on Climate Change from adopting the Copenhagen Accord. In response, in January 2010 Evo Morales, Bolivia's president, called for "the organization of the World People's Referendum on Climate Change," to "analyze and develop an action plan to advance the establishment of a Climate Justice Tribunal," and to "define strategies for action and mobilization to defend life from Climate Change and to defend the Rights of Mother Earth."[10] Climate writer Naomi Klein, whose ideas and role I consider later in this book, once called the Cochabamba process which produced the following points of unity "the most transformative and radical vision so far" (she is silent about these points of unity c. 2021).[11] The Cochabamba agreement was based on a "project for the Universal Declaration on the Rights of Mother Earth" which stated, "Mother Earth and all beings are entitled to all the inherent rights recognized in this Declaration,"[12] including:

- The right to live and to exist;
- The right to be respected;
- The right to regenerate its bio-capacity and to continue its vital cycles and processes free of human alteration;
- The right to maintain their identity and integrity as differentiated beings, self-regulated and interrelated;
- The right to water as the source of life;
- The right to clean air;
- The right to comprehensive health;
- The right to be free of contamination and pollution, free of toxic and radioactive waste;
- The right to be free of alterations or modifications of its genetic structure in a manner that threatens its integrity or vital and healthy functioning;
- The right to prompt and full restoration for violations to the rights acknowledged in this Declaration caused by human activities.

Forms of development which dump detritus in reservoirs and fill the air with waste and carbon dioxide, nuclear power, willful genetic tampering, infringement on cultural autonomy, and perhaps above all, industrial processes which spill out and over biomes' capacities for remediation would be banned. The document states:

> Under capitalism, Mother Earth is converted into a source of raw materials, and human beings into consumers and a means of production, into people that are seen as valuable only for what they own, and not for what they are. Capitalism requires a powerful military industry for its processes of accumulation and imposition of control over territories and natural resources, suppressing the resistance of the peoples. It is an imperialist system of colonization of the planet.

It argues for interlocking forms of restitution:

- Restore to developing countries the atmospheric space that is occupied by their greenhouse gas emissions. This implies the decolonization of the atmosphere through the reduction and absorption of their emissions;
- Assume the costs and technology transfer needs of developing countries arising from the loss of development opportunities due to living in a restricted atmospheric space;
- Assume responsibility for the hundreds of millions of people that will be forced to migrate due to the climate change caused by these countries, and eliminate their restrictive immigration policies, offering migrants a decent life with full human rights guarantees in their countries;
- Assume adaptation debt related to the impacts of climate change on developing countries by providing the means to prevent, minimize, and deal with damages arising from their excessive emissions;
- Honor these debts as part of a broader debt to Mother Earth by adopting and implementing the United Nations Universal Declaration on the Rights of Mother Earth.[13]

These are the planks of a southern platform for ecological revolution. They go well beyond the climate, to protecting Mother Earth as a whole, including creatures like insects which face huge threats.[14] They are the means to include the South in a just transition. And they are absent or underemphasized in most northern Green New Deals. And if any GND is a sweeping agenda for governing society, who is enfolded in any platform and who is excluded will define who counts and who ought to have their full humanity respected. The right to live is a mammoth demand that spills over from calls for eliminating ecological debt, into all the things that are needed in order to live – freedom from bombs and access to abundant food and shelter. Such breadth is not an obstacle to a wider process of change. Instead, the Cochabamba agreement is a reminder that to create a world where everyone counts, we need to set our sights to revolutionary horizons.

How can such radical demands connect with a GND in the core, which is right now on track to be – but need not be – elitist, separatist, and exclusionary? A GND need not lock in underdevelopment, ecologically unequal exchange, or the historical appropriation of carbon sinks. A GND need not rest on borders brimming with barbed wire and motion sensors. And a GND need not be a document aimed at bridling rather than breaking capitalism, rife with earnest evasive technocratic policy advice, a back-pocket policy guide for capitalists if the hydra of humanity keeps regrowing new heads faster than the forces of repression can decapitate it.

Unfortunately, far too much of the current back-and-forth on the GND does accept such compromises in the name of pragmatism. Or it accepts most of them, or some of them, or is silent on reparations, or most perniciously, makes reparations an afterthought, because (although sometimes this goes unsaid) debt repayments on the scale demanded by the Bolivian government are not "feasible." "Feasible" is the watchword of reaction and a talisman of oppression. It is the immaculate rhetoric of "objective" judgment and pragmatism, but is very often a cipher for identifying sacrificial victims for a great society that is only capacious enough to hold so many. This book aims to expand the scope of what is understood to be feasible. In so doing, I take as a point of departure four facts. One, existing GND proposals are broadly Eurocentric and rest on continued global inequality. Two, they are not ambitious enough to deal with the broader Earth-system's crisis. Three, many people in the core are ready for something more radical. Four, if the GND political debate and mobilization, which must be explicitly distinct from riding any specific legislation, is to be considered an agenda for governing and managing the future, it should ensure that the needs of the most oppressed and exploited are woven into its weft from the beginning.

Technical blueprints, discussions, drawing boards, and policy debates which foreclose climate debt, ecological debt, and reparations are not part of a transformative GND. They are more of the same. And any idea for a GND which does not bring "back the national question to the development agenda," in the words of agrarian researchers Sam Moyo, Paris Yeros, and Praveen Jha, simply cannot advance the struggle for liberation in its many colors and shades.[15]

(ENVIRONMENTALLY) UNEVEN EXCHANGE

This book takes a theoretical approach which I have tried to use in a way that can inform organizing in the imperialist core towards collective liberation. Its foundation, based on dependency theory, is that the world-system – global capitalism, in the common tongue – is inherently polarizing. In the core, or the wealthy states, in the words of Samir Amin, "development is characterised by the dominance of economic activity to satisfy mass consumer needs and the consequent demand for production goods." Furthermore, the core has a specific way of organizing access to what people need: "the power of the masses is enlisted in a 'social contract' which allows the establishment of a limited economic viability, at a national level."[16] Additionally, capitalism, or the historical process of accumulation on a world scale, is uneven over space and between nations. Wealth and well-being concentrate in the nations called cores, and poverty in places called peripheries and semi-peripheries.[17] It is a feature, and not a bug, of capitalism that wealth piles up amongst the few and poverty piles up amongst the many.

These are concepts for interpreting history. They are the appropriate lenses to see features of the historical landscape that other viewing devices do not bring into focus. They help us see that the capacity to produce wealth is not because core countries are exceptional other than in their use of violence against the periphery. These concepts allow us to think of the wealth of the core as based on the poverty of the periphery. In the words of Frantz Fanon, "In a very concrete way Europe has stuffed herself inordinately with the gold and raw materials of the colonial countries...there has flowed out for centuries toward that same Europe diamonds and oil, silk and cotton, wood and exotic products. Europe is literally the creation of the Third World."[18] Capitalism is tied to colonial and imperial value transfer: capital flowing in channels hewn by political violence, from Potosi's silver mines to Tunisia's phosphate pits to Congolese cobalt quarries, to India's jute plantations during the Raj to Haitian sugar plantations.[19]

In the common narrative, apparently fair or natural "market" forces, produced by supposedly free and equal trading partners, produce the price system. This sanitized story hides how political violence ensures that "consensual" prices preserve power in some places and poverty in others. The illusion that prices are fair or represent anything other than social power allows us to forget that prices and all the bestiary of macro-economic indicators and statistics which are involved in capitalist social engineering, from GDP to growth, are based on an intolerably violent enclosure: the atmosphere, as a dump for the waste products of capitalist industrialization, into which the West has dumped CO_2 for hundreds of years now. The price is now coming due, and those who pay it are first of all the world's poorest, not its richest. For one commodity after another, presented to northern consumers as the necessary trappings of civilization, life and land is lost in the South and profit piles up in the North. This is not just the case for agricultural crops but also the manufactures which are the source of so much northern profit. From the 1970s onwards, the poorer countries, the periphery and semi-periphery, industrialized but without coming close to closing South–North income differentials. The theory which explains differences in access to all the good things the world produces (or use values, in Marxist terms) is called unequal exchange: people in the periphery could produce the same widget at the same speed as someone in the core and be paid radically less for it. And even when we account for differences in productivity, wages in the South are far lower than they are in the North.

Environmentally unequal exchange (EUE) emphasizes that it is not only goods and labor that are enmeshed in unequal exchange. Cotton, for example, requires earth as well as labor. Peppers require fertilizers and in the arid climes of the southern Mediterranean, groundwater. And fertilizers can be safely sprayed from expensive machines in the North – although consumers still eat them – but in the South, it is human beings who spray them. So, it is not merely the natural wealth, the free gifts provided by nature, which are consumed unevenly along North/South lines. It is also the environment itself. Clean water ends up secured for northern populations by regulations which make it expensive to pollute and guarantee clean drinking water, although not everywhere: look at Flint, Michigan. Clean water is scarcer in the South, because regulations are not as strong – nor are governments, especially when they get too strong and carry out policies unfavorable to global capitalism and are tossed out by US coup d'états, like Evo Morales in Bolivia in 2019. The mines of Potosi produced tailings, and phosphate production produces phosphogypsum. Neighborhoods in the sublime Gabès seaside oasis, a unique emerald of the Tunisian south, are devastated by cancer to produce the phosphate products needed for industrial agriculture. The West has systematically shifted its dirtiest industrial plants to the semi-periphery and the periphery. Even in the core, the dirtiest waste processing is often sited close to Black neighborhoods. And centrally for our purposes, the cheap developmental paths enabled by coal and oil

– literally the highly concentrated products of the photosynthesis of the distant past, the sun's energy turned by time and pressure into nuggets and brews of black energy – poured gigatons of carbon dioxide into the Earth's atmosphere. What had long been the commonwealth of humanity was enclosed as a waste dump for the byproducts of burning, with the benefits going to the core. And for most of capitalist history, outside of the unusual 1945–1973 period, those profits went to a very narrow group of people in the core. Environmentally Uneven Exchange is the theory which helps us to make sense of this injustice, and to continue the claim of Fanon, "So when we hear the head of a European state declare with his hand on his heart that he must come to the aid of the poor underdeveloped peoples, we do not tremble with gratitude. Quite the contrary; we say to ourselves: 'It's a just reparation which will be paid to us.'"[20] Reparations during decolonization were about capital grants. Capital grants are still needed, but in the age of climate change, reparations must also repay ecological and climate debt.

STRUCTURE OF THE BOOK

Before structure, a short comment on terminology, and what this book is and is not. I freely switch between the set of terms Global North, core, imperialist and wealthy states, on the one hand, and Global South, periphery, subjugated and poorer states on the other. Core and periphery are the most exact for the reasons given above, and because geographically southern countries like Australia and New Zealand are also part of the core. However, the core/periphery couplet is probably unfamiliar, and it gets boring to use the same words.

On what the book is and isn't: This is not a theoretical treatise on planning for eco-socialist transition, although those books should be written. Nor is it a ready-made road map for local-level transitions in the 50 states, the imperial core, or the entire planet. Those are useful too. Nor is it a recipe book surveying every sector and every possible intervention. In particular, when it comes to industrialization and energy policy, it is meant to gesture at some possible problems with dominant ways of thinking as well as their possible blind spots. Above all, it is a set of extended interventions in an ongoing conversation, informed by the struggles and demands of the countryside and the Third World, and marked by my perception that Green New Deal discussions pay far too little attention to agriculture and climate debt, amongst other issues. Agriculture is indeed ever more urgent given the almost-certain source of Covid-19 in industrial agriculture, and the likelihood that agroecology could buffer such blights. Agriculture is, furthermore, central because it moves the discussion beyond climate reductionism, or reducing environmental crises merely to carbon emissions, to overall landscape management. I hope the book can be useful to organizers, activists, and all people interested in a better world.

The book is divided into two Parts. Part I, Capitalist Green Transitions, deals with proposed models for the green transition. It discusses those Green New Deals, or just green capitalists, which do not aim unequivocally for an eco-socialist world. It includes the following four chapters.

Chapter 1 looks at the proposals for Great Transitions from above: plans for financializing and commodifying nature and creating new kinds of investment vehicles from renewable infrastructure. Demographic fearmongering and engineering, and rising Malthusianism. And land use scenarios for capitalist or post-capitalist but still hierarchical "Great Transitions," which are attempting to push vegan diets, clearance of land for monocrop tree plantations, biofuels, and the other claptrap and paraphernalia of eco-modernization on the already devastated countrysides and Indigenous peoples of the planet.

Chapter 2 treats green modernization more theoretically and historically, by offering an account of how it came to be that boisterous enthusiasm for high-flying resource intensive technological fixes has soaked the cognitive atmosphere. It traces how major self-defined socialist currents have begun to repeat eco-modernist/modernization talking points, in the process undermining and disabling resistance to the capitalist agenda.

Chapter 3 focuses on the thorny issue of energy use, which in the West is the fulcrum of the transition to a post-carbon society. I discuss degrowth versus energy-cornucopian models based on green Keynesianism, and furthermore analyze the potential for a "new" uneven accumulation which is really the same old colonial division of energy usage baked into left-liberal green transition models. I then examine some of the odd assumptions and hidden ruptures with the precautionary principle baked into the majority of models for transition.

Chapter 4 examines the prominent progressive discussion of the Green New Deal, especially as it has departed from the Markey/Ocasio-Cortez legislation. It takes the opportunity to dispassionately read AOC's legislation, so that discussion can proceed based on what the text of the legislation states. It considers the merits and demerits of the most prominent volumes that are structuring discussion on the GND and considers whether they reject or affirm support for climate debt payments, how they deal with agriculture and land, and how they interact with emerging popular support in the West for systems change.

Part II, A People's Green New Deal, enfolds the three chapters which sketch out some elements of an eco-socialist strategy and an eco-socialist end-point.

Chapter 5 begins to set out some elements of what I consider important for a People's GND. It focuses on the political and social forces that would compose a People's GND coalition, primarily, but not only, in the North. It then discusses some critical sectors, including industrialization and manufacturing, design and architecture, and enormous shifts in transport to reduce overall energy use and increase pleasure and efficiency.

Chapter 6 focuses on agriculture, carbon drawdown, and land use. I argue that food sovereignty based on agroecological production methods should be at the center of a People's GND, not merely in the South, where the agrarian path to popular development is more obvious, but also in the North. The chapter moves from an account of pre-colonial agricultural systems to the costs incurred by industrializing US and world agricultural systems – on health, social reproduction, political freedom, and qualitative development. It then shows how a slightly more labor-intensive land use and agrarian path could rapidly draw down global CO_2 levels and increase farmers' and popular well-being, while improving the quality of food in the North and even the quantity of food in the South. It would also build protective dikes against animal-incubated viruses, ever more urgent to head off the next pandemic before it arrives.

Chapter 7 surveys the national question through several manifestations: sovereignty, climate debt, decolonization, and demilitarization. I argue that a peacetime economy and true respect for national sovereignty in the South are keystone elements of a GND in the North. I further argue the political and ecological grounds for decolonization and Land Back, not merely in the remaining "Third World" or abroad settler colonies or post-colonial economic apartheid models of South Africa or Palestine, but also in the imperial core, especially the US.

The Conclusion ties together the political threads from the rest of the chapters, suggests how I think change happens, and clarifies ways that internationalism can be built into environmental justice organizing from the outset.

The people of the valley did not conceive that such acts as they saw and felt much evidence of in their world – the permanent desolation of vast regions through release of radioactive or poisonous substances, the permanent genetic impairment from which they suffered most directly in the form of sterility, stillbirth, and congenital disease – had not been deliberate. In their view, human beings did not do things accidentally. Accidents happened to *people, but what people did they were responsible for. So these things human beings had done to the world must have been deliberate and conscious acts of evil, serving the purposes of wrong understanding, fear, and greed. The people who had done these things had done wrong mindfully. They had had their heads on wrong.*

Always Coming Home, Ursula Le Guin

PART I
CAPITALIST GREEN TRANSITIONS

The first part of the book analyzes the ruling class agendas, especially the Great Transition literature and its theoretical and historical foundations. It also dissects certain arguments which support that transition, and comments on the political strategy accompanying progressive programs which strongly differ from the Great Transition agenda in many respects, but may offer it insufficient resistance in others.

If we want to go somewhere and people more powerful than us do not want us to get there and have plans for where they want to go, it is urgent to understand the nature of those plans. They offer a contour map of the political battlefield and clarify the means and aims of the opposition. For that reason, the entire first chapter shows what the ruling class intends to do, how it intends to do it, what the costs of doing it will be, who will pay those costs, and what those plans will do to the countryside and its people: the central subject of social transformation and the well-spring of value.

In paying attention to land, I zero in on a topic commonly ignored in writing on capitalist, green capitalist, green social democratic, and sometimes eco-socialist transitions. The focus on land is important on its own terms. However, it is critical in showing how much the outwardly capitalist Great Transitions and the avowedly socialist eco-modernist literatures, alongside the compromises of some social democratic Green New Dealers, create a technological alignment around what to do with the countryside. While these thinkers often, but not always, differ on the nature of rural social relations – I say often, because some accept purging pastoralists, simply eliminating that rural class – their embrace of the capitalist technological agenda leaves such plans acutely vulnerable to co-option. Increased industrialization of agriculture, the removal of labor, even depopulation and use of the land for biofuels and nature reserves are common themes, converging on a radical separation of people from nature.

Because technology is central to all such plans, I have devoted a chapter to understanding the stakes of different conceptions of technology, and pushed back against the tall tale of the categorical neutrality of all technologies.

Finally, some may wonder at my inclusion of social democratic Green New Deals amongst the capitalist Great Transitions. First, the chapter had to go somewhere. More substantively, for reasons the chapter explains, advocating for domestic social democracy while failing to seriously engage with the colonial legacy and neo-colonial and imperial present is an agenda for continuing that present, now and in the future. In conceding space for capitalism to continue, because profits have to come from somewhere, the social democratic agenda undermines its own aims, and forgets how social democracy was achieved: never through parliamentary struggle, and always in the presence of far more radical threats domestically and especially on the Eurasian landmass. For those who want an eco-socialist world, it will not be helpful to aim at horizons well short of that in the name of realism. Such a strategy is not realistic at all.

1

Great Transition – or Fortress Eco-Nationalism?

Concern about the climate crisis has become overwhelming in the Global North. It is no longer the isolated worry of climate scientists, forward-thinking investors, coastal real estate proprietors, or "environmental" activists. From financiers at Davos to McKinsey Consulting, it is everywhere, as the business press and mainstream media emit blizzards of policy papers, communiques, and financial forecasts.[1] Why? In part, climate change is now unmissable. Eighteen-degree Celsius days in January in New England. Catastrophic floods and hellish heatwaves across the US Great Plains and Paris. Climate change is no longer in the future tense. It is today. The media is forced to pay attention as people understand more clearly what is happening and why it is happening. Yet most media outlets are owned by big corporations that are part of the capitalist system responsible for climate change.[2] So the speed and simultaneity with which the ruling class has changed its tune, from decades of denialism to widely touted tabulations like that of the World Economic Forum "that $44 trillion of economic value generation…is moderately or highly dependent on nature and its services and is… exposed to nature loss," ought to make anyone curious.[3]

For climate change does not dictate any specific political response. This chapter considers the rising chorus for "emergency change" from those with their hands on the pulleys and levers of the machine, in order to understand how top-down plans seek to maintain exclusion and exploitation in the world-system. Not the 1 percent but more like the .01 percent, who have seen incomes and net worth scrape the sky during the neoliberal age. They wish to preserve an imperial way of life and the planetary resource base upon which to live it. Their hands, pens, and plans are not idle. Amidst rising awareness of the capitalism–climate nexus, it is only natural that the ruling class would seek to avert a climate crisis which could imperil their power, to displace blame from fossil capitalism to a faceless and structureless humanity, and to make the poor pay the costs of transition.

There is not a whit of concern here for the well-being of the working class, or more than a few grains of worry for non-human life. If that were the case, they would have taken action on climate change 30 years ago. Rather, a constellation of forward-looking right-wing and liberal forces plan for what the global sociologist Philip McMichael calls "managing the future."[4] These range from the Australian Breakthrough Institute – a doppelganger of the Ted Nordhaus think-tank with the same name based in California – the Energy Transitions Commission, and on to figures often associated with the progressive left. Such proposals share a number of leitmotifs: corporate–community and corporate–state/business–state partnerships, warmth to the national security sector, a Promethean enthusiasm for "new tech," recasting models to integrate variables like climactic risk and enfolding them into investment horizons, and a greened US military. A large number are converging on what the economist Daniela Gabor calls the Wall Street Consensus, reorganizing "development interventions around selling development finance to the market…escort[ing] capital" into bonds, and melting and reforging Third World governments as the "de-risking state," relegating risks onto states and their treasuries and budgets as sovereign representatives of the people and removing those risks from the mountains of capital which such plans seek to mobilize.[5]

GREEN SOCIAL CONTROL

Green Social Control aims to preserve the essence of capitalism while shifting to a greener model in order to sidestep the worst consequences of the climate crisis. Because there is no capitalism that exists apart from the violent hand of the state, such plans emphasize the national security sector.

Drawing borders around the people, denizens, or citizens of the nation means exclusion.[6] Structural exclusion, from popular development. Ideological exclusion, from concern or care. Physical exclusion, through hardened steel, concrete, razor-wire and motion-sensor kitted-out borders and militarized maritime zones. And temporal exclusion, in the erasure of a past of colonial looting and atmospheric enclosure. Such plans are engineered by the best and the brightest, and we have to take them at their precise word. What they say or do not say about the Third World is what they mean. And because the US is an empire, sitting astride tremendous flows of energy and capital, silences speak. A decision to avoid reparations is a blueprint for world management in which imperial loot remains in the North.

Part of the right's Great Transition is to match up existing assets such as working-class pensions, or the public resources of the people in the form of the state budget, with new tech to harness and commodify the sun and the wind through solar power and wind energy. Greening the US military is central. Putting prices on "eco-system" services, including a potentially serf-like labor-intensive CO_2 drawdown program, is a third pillar of their programs. And because the framework of threat is paramount in securing the consent of the core citizenry to their programs, the alarm bell of emergency blares daily (in this case accurately) to frighten people into accepting any plausible program to preserve a livable Earth.

A CLIMATE OF EMERGENCY AND RISING RISK

How we describe a problem helps mold how we will address it. Notions of rising emergency and climate-related risk hide something pointed out by the German philosopher Walter Benjamin: "The tradition of the oppressed teaches us that the 'emergency situation' in which we live is the rule."[7] Lawlessness, disorder, threat, human insecurity, privation, deprivation, and suffering are the rule for most of humanity. Endless war, primitive accumulation, colonial genocide, neo-colonial depopulation and state-shattering are permanent emergency.

The rhetoric of risk is not a recognition of the emergencies that batter the lives of the world's majority population. The newly noticed threats, about which the core's governments have blithely done nothing for decades, are to long-term accumulation: social stress in the South leading to drought and mass migrations, and now a chance of runaway warming. The World Economic Forum, for example, slots risks into "physical," "regulatory and legal," "market," and "reputational": perils to business. Such parochialism is risk and emergency from the very particular viewpoint of Western monopolies.[8] If open and democratic debate about the distribution of wealth can be muffled under the smothering blanket of emergency, those who own the blanket and insist on asphyxiating discussion can preserve existing distributions of wealth.

Other uses of emergency are the canary in the coalmine for an acceleration of the worst of current fortress nationalist regimes, those countries which are increasingly militarizing their littorals and southern borders to stem immigration from countries which they have often made unlivable. Worry about mass migrations is worry about whether the cloistered settler-states and the European capitalist heartland will remain safe. What will Australia do if Indonesia is inundated and monsoons fail across rural India? Mexico, the Caribbean, and the Central American isthmus are even closer to the US, and large-scale Central American migration is already taking place, as hurricanes and brutal heat smash vulnerable farmers in a strip of land long concussed by US interventions.[9] Emergency and panic about amorphous threats to life in the Third World have always justified illegal US aggression, much as neo-colonial environmentalists fretted about the forests of the Bolivian Amazon in 2019, to try to blacken the globally unparalleled ecological advocacy of the Indigenous-led left political party MAS.[10] Furthermore, the war always boomerangs back to where it started through settler-colonial frontier wars.[11] Suspending law, civil liberties, democratic deliberation, and struggles for liberation amidst warnings of a falling sky or "terror" have been common in a post-9/11 world, the pretexts for deepening hierarchies of class and nation.[12]

Against a renewed imperialist urge to generalize the "state of exception" through new regimes of state–private accumulation, greened and global, we can process the sudden cacophonous concern for climate change amongst the planet's wealthy and their court intellectuals. Take some examples. A recent Brookings Institute report discusses rising "droughts, fire" and natural disasters. These plagues translate to "risks to economies and livelihoods." Furthermore, the "*current growth path*" pressures water, land, and biodiversity, causing "accelerated loss of natural capital."[13] Words like economies and livelihoods are kaleidoscopic. They hint at human needs for food, shelter, and the good life and conflate them with the well-being of the economy. In this discourse, it is simply assumed that the health of "the economy" leads to the health of concrete individual human beings. Their history is important: the blurry notion of livelihoods emerged in force at the Rio+20 Summit, which marked 20 years after the original Rio Earth Summit which set in motion world climate negotiations. At Rio+20, mandates for "sustainable development" of the world's resources bloomed, alongside sterile calls for a "green economy," and were adopted by consensus at the United Nations General Assembly.[14] Yet in an imperialist world, that which all states agree upon as a program cannot have any emancipatory content. Neither economy nor livelihoods implies any particular distributional pattern: "Class, not as an institutional context variable, but as a relational concept, is absent from the discourse of livelihoods" argues sociologist Bridget O'Laughlin, which is why it studs such reports so frequently.[15] They do not wish to question the existing maldistribution of wealth.

Take another example: the Australian Breakthrough Institute, in a May 2019 report, *Existential climate-related security risk: A scenario approach*, offers a perspective from David Spratt, the institute's in-house research director, alongside Ian Dunlop, a former international coal, oil, and gas industry executive, and the chief executive of the Australian Institute of Company Directors.[16] The foreword by Admiral Chris Barrie calls for "strong, determined leadership in government, in business and in our communities" to combat the cataclysm of climate change (akin to kindred blurring devices like economies and livelihoods, "communities" fuses interests of the rich and poor within nations and is silent on diverging interests between nations). The report warns of "existential risk," and urges "we must take every possible step to avoid it." How? Under a scenario better than the current emissions trajectory, with an emissions peak at 2030 and an 80 percent decline by 2050, "Deadly heat conditions persist for more than 100 days per year in West Africa, tropical South America, the Middle East and South-East Asia," displacing one billion people. Water availability would decrease in the dry tropics and subtropics, parching two billion. Agriculture would become "nonviable in the dry subtropics." As watersheds and temperature bands within which humans millennially flourished drift north, social structures erected on a more or less stable climactic bedrock would shudder. We could "see class warfare" as the wealthiest "pull away from the rest," shattered international trade flows, a retreat into autarky, collapses in economic interchange, and the inability of governments to govern.

Such national security bureaucrats and military-defense intellectuals know climactic shifts threaten to shift the planetary ecology sufficiently severely to evaporate the gossamer logistical flows and political scaffoldings of global capitalism. They write, think, and plan accordingly. These reports are not the handicraft of melioristic humanitarians, but diamond-hearted strategists dumbfounded at the sluggishness with which most of their fellow ruling-class and governments have acted.

To avoid the massive human migrations such a scenario would entail, the report calls for "a zero-emissions industrial system [to] set in train the restoration of a safe climate." The predictable analogy they draw upon is World War II, the crucible of military Keynesianism and worldwide neo-colonialism. The report highlights the "national security sector'[s]" potential role in a vast mobilization of resources. The role of the military in a peacetime transition is unclear, although one chapter of World War II less discussed by its refurbishers, right and left, is the role of the military in suppression of labor unrest, or union agreements to avoid independent labor action.[17] Furthermore, they propose enlisting the national security sector to assist resource and labor mobilization. Such help could mean anything, but gestures at forced labor recruitment. To carry out this program, they imagine a new social consensus, based on a triangular relationship between Parliament and "corporate leaders," and community. The former will provide wisdom and "leadership," the latter will provide support and internalize and normalize in "everyday discussion" the logical conclusions of emergency planning.

What does that look like? A wholesale renovation of the technological and institutional infrastructure and architecture of fossil capitalism, the new product a lot of the same old capitalism but covered with honeyed rhetoric of sustainability, community, livelihoods, and resilience, so it can be smoothly swallowed by those who might balk at climate crisis as an occasion to socialize risk and privatize profit. There is a logical and political correspondence between the language and strategies they use to address the worsening emergency for humanity, and their positions in society: elite management, links to the most polluting forms of fossil capitalism, and the national security state. We would expect them to frame the problem, identify stakeholders, and propose solutions in ways that do not disturb the patterns of power and powerlessness from which they speak, and which they wish to preserve.

THE RETURN OF MALTHUS

One aspect of the "emergency" discourse that is slightly new, albeit a recurring reflex in the course of capitalist social management, is Malthusian thought. An economist, Thomas Malthus floated the folly that the poor's food consumption would rise faster than agricultural production, and asserted that the poor had no right to poor relief and no capacity to control their numbers. Giving them alms would undermine one of the checks on their numbers, producing an intractable biological-natural problem: food supply supposedly increased arithmetically, whereas population could increase exponentially.[18] These were fabulations heaped on misinterpretations heaped on nonsense.[19] Famines tend to occur due to decisions to deprive certain populations of the power to access food, not absolute lack of food.[20] Contemporary worries about food production are founded on similar social science fiction, for far more food is produced than humans currently need, much of the food that's produced is nutritionally hollow, and far more food could be produced than is produced currently.[21] Claiming otherwise is a Trojan horse for the entry of capitalist logic into dissident thought. The aim is to prevent the conscious and collectively decided use of what is produced to feed those who need it most: the working class, whose labor directly produces almost everything anyway.

The re-emergence of Malthusianism marks a moment when wealth-holders and their stable of captive thinkers are figuring out how to hoard resources for themselves and to justify the imposition of misery elsewhere on the planet. To take one example, a mass gathering of world scientists recently penned an open letter claiming, "the world population must be stabilized – and, ideally, gradually reduced – within a framework that ensures social integrity."[22] Or take the organization Population Matters, on whose board sit prominent public personalities Jane Goodall and David Attenborough, who recently appeared in a video with Greta Thunberg. It states, "Biodiversity loss, climate change, pollution, deforestation, water and food shortage – these are all exacerbated by our huge and ever-increasing numbers. Our impact on the environment is a product of our consumption *and* our numbers. We must address both."[23] Such a statement may jibe with an arithmetical common sense built on the axiom that humans necessarily impact the environment, producing the equation: More humans → more impact. In truth, world population is largely irrelevant to the climate change question. CO_2 emissions are primarily rooted in capital accumulation, not in too many babies, and emissions are the responsibility of the wealthiest in the wealthy North. Blaming an undifferentiated world population, a demographic phantasm, sidesteps the difference between a Bill Gates and a Bangladeshi small farmer. It also sidesteps that humanity is already well over the carbon budget, and those who continue to spew out more CO_2 than the earth-system can metabolize are overwhelmingly the less numerous rich rather than the more numerous poor. The incantation of the Malthusian specter is also a coded call for more brutal forms of population control. The Bill and Melinda Gates Foundation has been front-and-center among the capitalist foundations agitating for African population suppression.[24] The demographic modelers are aware, as are the Gates and other foundations, that population growth primarily occurs amongst the planetary poor. In response, "populationist strategies focus on targeting the number of humans on the planet (demo-populationism), the containment and shaping of populations in relation to particular spaces (geo-populationism)," which implies severe intervention into the reproductive lives of women.[25]

Furthermore, from the perspective of the ruling class, the growth of "relative surplus populations," or those under- or unemployed, provides a second edge, no longer merely a blade with which to cut into working-class wages, even below the reproduction cost of labor.[26] And while surplus populations always represent potentially insurgent and anti-systemic claims to development, under capitalist imperialism they were a useful cog in the social machine in grinding down wages. But under contemporary capitalism, or whatever comes next, the poor sit atop savannahs, forests, and prairies, which can be used for biofuels, and carbon farming through ghost forests of deciduous and coniferous monocultures, each molecule of carbon absorbed through a new tree representing another molecule which can be emitted from incinerated oil, and another dollar of ExxonMobil's assets secured from environmentally precipitated expropriation.

THE MILITARY AND THE CLIMATE

Capitalism and militarism go hand-in-glove. The age of gunboat imperialism never ended. The need for a robust and nimble military to secure accumulation on a world scale, including through the erasure of Arab, African, Asian, and Indigenous life, has been an unending feature of capitalism.[27] From a technical-logistical perspective, as the closed fist of imperialism, the military has long been concerned with its operational capacities in the face of climate change. From one perspective, such concerns are ludicrous. One cannot deal with climate change while preserving the Pentagon system, one of the planet's primary polluters.[28] Furthermore, such concerns should be taken gingerly in other respects. Department of Defense warnings of "terrorism" have long been pretexts for force projection, as with attempts to link Saddam Hussein to Al-Qaeda – the former of whom was a former US client, and the latter which grew out of the US arming of the *mujahideen* through Pakistani middlemen, which did not prevent the US from annihilating Afghanistan and Iraq.

Many DoD responses to climate change are alibis to advance enhanced military presence domestically and abroad. Still, other concerns include the actual fighting and fueling capacity of the military itself, and the potential of increased population movements, which can only be met by a military response. Again, the social containment and population management framework sets the terms for the response, with words like "risk" and "threat" threaded through the relevant documents.

The Biden administration is already interested in "Invest[ing] in the climate resilience of our military bases and critical security infrastructure across the U.S. and around the world."[29] And such thinking has long permeated the strategic thinking of the US's military arm of accumulation. Consider a 2015 Department of Defense response to a Congressional Inquiry, framing climate change as a "threat multiplier," causing cascades of refugees as governments grapple with "meet[ing] the basic needs of their populations." Amidst accelerated climate disaster, it suggests the DoD may need to carry out "humanitarian assistance and disaster relief" abroad and Defense Support of Civil Authorities at home, which will "likely rise as cities expand to encompass the majority of the global population." Sea-level rise, the report warns, may imperil ports, while the opening of the Arctic amidst evaporating sea ice is not a cataclysm but an opportunity – "greater access for…shipping, resource exploration and extraction, and military activities."[30] A January 2019 *Report on Effects of a Changing Climate to the Department of Defense* notes that between 50 and 100 percent of Air Force, Army, and Navy installations are at risk of recurrent flooding, drought, and wildfires. One base in Langley is experiencing "more frequent and severe… flooding."[31] The DoD takes resilience planning extremely seriously, deploying measures from increased minimum height-above-sea-level to adaptive planning measures, to boosting the capacity of servers and back-up generators to work in hotter climates.[32]

The Elizabeth Warren-authored Department of Defense Climate Resiliency and Readiness Act is the most comprehensive statement of how to change the military to deal with climate change. As Warren announced the need to gird for an age of zero-carbon, "We don't have to choose between a green military and an effective one. My energy and climate resiliency plan will improve our service members' readiness and safety."[33] By definition, such planning does not question what the military does or its existence. The Act details the Warren vision, and centers around the baggy euphemism, "net zero energy," which usually means continued emissions:

> A reduction in overall energy use, maximization of energy efficiency, implementation and use of energy recovery and co-generation capabilities at each installation, and an offset of the remaining demand for energy with production of energy from onsite renewable energy sources

at such installation, such that such installation produces as much energy as it uses over the course of a year.[34]

The Act demands development and application of the readiness-and-resilience protocols outlined in earlier Pentagon communiques. Rehearsing post-World War II use of military spending as a Keynesian multiplier and central political mechanism for a semi-command economy under liberal democracy, the Act suggests subcontractors consider whether the industrial plants to which the Pentagon submits weapons and materiel orders produce their own renewable energy or are in violation of environmental legislation. The DoD is to become a hub and laboratory for R&D on "hybrid microgrid systems and electric grid energy storage," and as part of that, "local generation of zero-carbon fuels," or biofuels.[35] Perhaps most tellingly, shifting internal Pentagon R&D towards renewable energy meshes with private-sector plots to shift the costs of further advances in alternative and purportedly renewable energy systems onto the public.

Finally, the program calls for US states, "Indian tribes," "regional entities and regulators," "institutions of higher education, *including historically Black colleges or universities and other minority-serving institutions*," and "private sector entities" to do the research with appropriations rising to $250,000,000 by 2026.[36] The amounts are picayune by Pentagon standards, but they reflect efforts to envelop in the multicolor cloak of diversity some meagre, patchy-green shifts in the US economy and in core institutions like the Pentagon system. The effort is not merely symbolic. It could establish financial interests for representatives and leaders of communities of color into an elite Great Transition. Such diversity-from-above is part and parcel of the need to acknowledge, incorporate, placate, and vacate the victories of grassroots movements which have succeeded in forcing token recognition of historical grievances related to Black slave labor and conquest of First Peoples' lands, the original accumulation upon which US capitalism rests. Under the great green transition, the Pentagon system is going nowhere.

We now turn to the broader set of interests in which the Pentagon system is enmeshed, and alongside which the "Great Transition" framework is shifting towards green accumulation.

A NEW ROUND OF ACCUMULATION

Green Transition models plan to mobilize public finance to commodify the free gifts of nature. They go beyond current thefts, through the silent purloining and despoliation which the jargon names as "externalities," turning those natural gifts directly into a new asset class. This is no new trick. The notion of climate debt recognizes that North Atlantic capitalism enclosed the atmosphere as a dump for its waste eons ago. What is slightly new is how epochal shifts in the natural environment due to capitalist-induced destabilization are pushing the ruling class and its scribes into adapting the apparatus of accumulation and hierarchy. Some of this planning is visible in statements from financial engineers and fund managers. After decades of hedging, hemming, and hawing, many such managers have started to see how shifts in natural processes may push them to shift investments. The British banking sector is one of them. The Economist Intelligence Unit estimates risks to global assets by century's end could total $43 trillion, and some consider the bell may be tolling for international oil companies, which may be "negatively revalued."[37]

But capitalism destroys in order to create. The foreclosure of some fields of investment means prising open others. For the burgeoning "Great Transition," several arenas are key. First, infrastructure, enfolding the gargantuan electricity and power sectors – the core of industrialization, and a central arena of interest for the Wall Street Consensus. Second, perhaps much more ominous, is the reboot of wrecked REDD and REDD+ agendas. The acronym stands for reducing the emissions from deforestation and forest degradation. The + adds in conserving stocks of forest carbon, sustainable forest management, and the enhancement of stocks of forest carbon. However, these programs often wove deciduous woodlands and tropical rainforests into the price nexus, financializing nature through carbon markets which put a price tag on "eco-system services."

Now consider three keystone "Great Transition" plans which sketch out the world they wish to see: from the Brookings Institution, from the Climate Finance Leadership Initiative, and from planning impresario and jack-of-all-capitalist-trades wordsmith Jeremy Rifkin. The Brookings report focuses on a new and more "sustainable" infrastructure which will "literally lay the foundation for a new and better growth trajectory and achieving the ambitious development goals."[38] Werner Hoyer of the European Investment Bank adds the EIB should take on the mantle of a dedicated climate bank, targeting "low-carbon investments," with one trillion Euro over the next decade. As he adds, "the climate crisis...represents an opportunity, because funding for new green infrastructure will create jobs, spur economic growth, and reduce the air pollution that is choking the world's cities." A substantial swathe of the new investments should be public, not private. He notes, "we are essentially paying to promote climate change and air pollution through fossil-fuel subsidies," a phrase which often refers to either the military-industrial infrastructure which allegedly protects fuel supplies, or alternatively the "externalities" of oil and coal incineration. (In fact, oil-fueled fossil capitalism has been a mechanism for maintaining the dollar as the world's fiat currency, and an element of accumulation on a world scale. It has also been part of the energetically "cheap" development path pioneered by the Western industrial capitalist states.) Those public funds ought to be "redirected toward investments in electric vehicles and other game-changing technologies that will drive the green transition."[39] Left unmentioned is the ownership of such vehicles and other technologies – the people of the planet, or the ruling class.

The most telling because most openly cynical of Great Transition initiatives is the Climate Finance Leadership Initiative (CLFI), the brainchild of United Nations Secretary-General António Guterres. He asked Michael R. Bloomberg, billionaire former mayor of New York City and UN Special Envoy for Climate Action, to chart the path towards a wide mobilization of private finance to respond to the crisis. Executives of seven key institutions – Allianz Global Investors, AXA, Enel, Goldman Sachs, HSBC, Japan's Government Pension Investment Fund (GPIF), and Macquarie – joined Bloomberg in CLFI. Its agenda? Private finance will enter zones of "availability of risk-adjusted returns matched to investor requirements." Where returns are "uncertain or higher-risk," however, "the private sector may work with public finance institutions and their blended finance, risk-sharing, and pipeline development tools." In particular, CLFI notes, this is because of uncertainty even in "core" markets, where amidst policy reversals, there are not "stable revenue models." How to ensure stability and suppress risk?

> Public budgets can continue to play a central role in clean energy deployment by guaranteeing revenues, especially in new markets and for newer technologies. Revenue security plays a decisive role in making clean technologies more attractive than carbon-intensive alternatives and providing investors with the confidence to deploy capital over longer periods.[40]

Furthermore, especially in "new markets," or the global periphery, the build-out of renewables may be more complicated. To reduce those political complexities, "Governments can also review permitting and litigation rules to help minimize project delivery times and avoid cancellations" – permitting and litigation are the ways in which afflicted Indigenous or poor populations can judicially challenge or forestall the installation of high-impact technology. Such rules, rendered in the antiseptic accountant's prose, will remove the mechanisms which peripheral populations can use to force their governments to subject developmental infrastructure investment to calculations which go beyond dollar value. Meanwhile, the CLFI clarifies it is government's job to guarantee "revenue security": the constant flow of value. Some mechanisms for that include "development finance institutions," to open "new markets" by crafting the appropriate investment climate. North and South differ, partially because extensive guaranteed tariffs in the North have often helped push the costs of solar and wind for new installations below the cost of the new hydrocarbon investments and energy systems. Furthermore, there are well-established mechanisms for pricing and payment of energy costs, which guarantees long-term stability of revenue flows. In the South ("emerging markets"), however, development finance institutions:

> can leverage private investment through risk-sharing tools, such as guarantees and political risk insurance, and their ability to source and coordinate catalytic finance from donors and third parties....Policy stability is also critical. Reversals or renegotiations of [Power Purchasing Agreements], tax incentives, or other agreements – particularly in the early stages of market development – can have a long-lasting negative impact on future investor interest.[41]

Political risk insurance guarantees against expropriation or nationalist governments renegotiating the rates their people pay for electricity. States will become permanently liable in international tribunals for internal redistributive action taken by future governments, placing them under the jurisdiction of courts wherein corporations can sue states for slicing away at their profits. The report notes that one of the risks to be buffered and hedged, and that must be considered by private finance, are "policy risks, such as expropriation [and] sovereign breach of contract."[42] The new green financing mechanisms rest on the northern capitalist enclosure and evaporation of sovereignty, resubmitting the South to colonization through financial chicanery, in turn relying on a legacy of colonial and neo-colonial underdevelopment and de-development, such that the North is the storehouse of financial and technical capacity for a southern shift to renewable energy. In this context, it ought to be highlighted that climate negotiations, such as at Copenhagen in 2009, were at perpetual loggerheads over northern unwillingness to imagine technology transfer, or the use of northern-developed technology in the South, on non-imperialist terms – which could include co-development, open access, or other ways of sharing knowledge while increasing southern capacities, rather than heightening southern dependencies.[43]

Furthermore, the CLIF supports the giveaways of carbon markets, with an assist from public finance "in markets and sectors where the introduction of fully-fledged carbon markets is not yet viable or when carbon market prices are too volatile to support long-term investment." The introduction of carbon markets suggests carbon absorption, mitigation, or reduction will be part of the Great Transition. Such calls invert polluter-pays principles, paying polluters not to pollute. Public underpinning of private investment also takes the form of the Pentagon–industrial state–private socialization of risks and costs and privatization and enclosure of social wealth by supporting the "commercial viability of earlier-stage, low-carbon technologies through funding for research, development, and demonstration."[44]

The proposal is similar to Jeremy Rifkin's 2019 *The Green New Deal*, a mix of technobabble and techno-baubles like the information economy, the internet of things, the Fourth Industrial Revolution, and the state-supported morphing of infrastructure to support a new social-industrial system. For Rifkin, the social basis for this new order is "a healthy social-market economy that brings together government, industry, and civil society at every level with the appropriate mix of public capital, private capital, and social capital."[45] The framework protects the prospective profits of private capital, which Rifkin and the CLIF make clear is incapable of taking the lead in clean-industrial transformation. The project's core is swaddled in the soft promise of a "social market economy," a 2000s-era technocrat's invention which gestures at softening the most abrasive impacts of capitalism through a notion that the market might be lightly "embedded" in society.

Rifkin argues for pension-based capital as the vanguard of industrial transformation (bracketing that pensions are small outside of the imperialist core).[46] The other sources of monies are small wealth taxes, and a miniscule 4 percent redirection of US Department of Defense spending. The call to reinvest pension funds, alongside other sources of private capital and government spending, is a call to use the remainder of social wealth to invest in a Second New Deal. This is a bald call for state-guided resurrection and transmogrification of the US industrial plant and its infrastructure.

Social guarantees would protect rights of workers to "organize and collectively bargain," whereas pension beneficiaries would enter the socio-political field of the new social-market economy as "little capitalists," fusing the interests of blue-collar and white-collar workers with magnates.[47] A sunny and pristine we-are-all-munchkin-scale-capitalists vision, not very different from that of Ed Miliband of the British Labour Party and Caroline Lucas of the Green Party, who call for "an industrial strategy seeking not just to mitigate climate change but to unlock new opportunities for investment and innovation, tackle inequality, improve quality of life and deliver an environmentally sustainable economy."[48] Nor very different from the rising star of the new industrial policy framework, Mariana Mazzucato: "The Green New Deal must have aspirations far beyond just mitigating climate change, and must be focused on new opportunities for investment and innovation – it must include finding clarity and courage in the policy arena, unlocking investment in the business sector, and supporting workers to acquire new skills."[49] Multiple figures here advocate a prominent role in private finance for the future "green transition," the state leading while protecting future profits and absorbing "the physical risks of extreme climate events or global pandemics that would strand infrastructure assets."[50] More importantly, this is the basis for a post-carbon social pact, weaving together the industrial and service working class in the core states with the state and private corporations.

GREAT TRANSITIONS AND LAND

CLIF and the Rifkin GND are largely silent on agriculture. But land use is key for their plans, as large capital wants to alchemize arenas now simply poisoned and pulverized with "externalities" into new assets and new use-values, from organic CO_2 pumps to capitalized nature. To put a green spin on such world-scale ecological-financial engineering, hand-of-god fantasies about Half Earth are gaining ground, imagining half of "Nature" immured from hordes of polluting souls – the annoying human components of the "Anthropocene" – through wide-scale conversion of human-inhabited areas to CO_2-drawdown farms and fortress conservation in which the wealthy world can luxuriate through luxe safaris, while clustering humans into the other half.[51]

This would be a bad idea even were it to be carried out by liberal humanists who carry in their intellectual satchel endless unseen and unexamined colonial baggage. But big finance and meta-social engineers are not letting melioristic if slightly misanthropic liberal biologists decide what should be done with land which is not remotely "empty" but filled with biodiversity-preserving Indigenous people, pastoralists, and peasants. They have their own ideas and bankroll their own platforms upon which to publish them. Mark Lynas, for example, on the website of the Gates-funded Cornell Alliance for Science, demands that "we" intensify agriculture, and corral off half of Earth to stave off starvation.[52] As we shall see, food production is not an issue and will not be an issue, so arguing for further industrialization of agriculture might be the Gates agenda, but has no humanist-scientific grounding – nor does fortress conservation itself.[53]

Elsewhere, the Nature Conservancy and its partners posit a "financing gap" of close to a trillion dollars to put an appropriate value on ecosystem services, including biodiversity – a new arena of commodification for capital currently sitting in low or negative return bonds, and another state "crowding in" of investment.[54] Another use of nature is biofuels.

The means to empty out such lands are Half-Earth fantasies and primitive accumulation of pastoral and grazing lands. By now, it has become faddish to focus on the merits of vegetarianism and veganism for climate change and health alike. Tellingly, then, when it comes to agriculture, these ruling-class institutions reduce the ecological and nutritional Armageddon of industrialized agriculture to methane and other CO_2 and CO_2-equivalent emissions from large ungulates. As with all colonial atrocity, this civilizing mission is said to be for love of humanity. They hide behind the scientific parapet of the EAT-Lancet report, and demand "plant or culture-based meats," which now rely on tremendous inputs of energy, and according to lifecycle assessments, may be more carbon-intensive than cows.[55] Such Frankensteinian lab-meats, grown in mammoth vats or bio-reactors, do have the merit of something rustic and landrace cows, goats, bison, and sheep do not: they are patented, the property of private corporations, and have big venture capital behind them clamoring to enclose and commodify what is now primarily a petty commodity or peasant-based form of production and turn it into a new high-tech monopoly. They also claim the newly liberated land could be reserved for "increased forest cover and biodiversity or if needed to bioenergy production." However, as shown below, the technology is hardly neutral. If the problem is industrial animal agriculture, why not simply produce meat using non-industrial methods? And if the problem is we can't produce enough, we ought to match up consumption with capacity for metabolically restorative forms of animal husbandry, rather than imagine that US leftists can simply "nationalize" venture capitalist enterprises, which will gobble up ever more energy at a moment when the crucial issue is suppressing, not boosting, aggregate energy use.

Furthermore, EAT-Lancet calls for potential "afforestation": putting forests where they have historically not existed. Or reforestation, referring to restoration of felled forests.[56] They sidestep that, historically, afforestation has been the green veil behind colonial advance.[57] Even now, planting trash-tree monocrops that may rampage over existing ecologies, severely damaging water tables, is a mechanism of colonial destruction of ecosystems, as in Palestine.[58] Additionally, the Edenic vistas of the past which justify such massive socio-ecological engineering may be largely imaginary.[59] Europe may never have been the land of blanket forests that lurks behind ideas of large-scale reforestation to put landscapes back to their "natural" state. Finally, these damages are not at all hypothetical: converting grazed grasslands into monoculture timber plantations has pushed the alabaster Indian Gaolao cattle breed close to extinction, destroying gossamer ecosystems and starving households of incomes.[60]

Once again, we should consider where bad ideas come from. They are not just the fruit of bad thinking. We should follow the money. The report was backstopped by EAT, a non-profit founded by the Stordalen Foundation, funded by Norwegian hotel billionaires, and the Wellcome Trust, on whose Board of Governors sit *inter alia* Eliza Manningham-Buller, president of the imperial think-tank Chatham House, and Richard Gillingwater, the chair of the renewables conglomerate SEE.[61]

Others like the World Business Council for Sustainable Development are jumping on the regenerative-agriculture and "vegan" tractor, ensuring it does not enter red-marked domains such as agrarian reform on a global scale. That group, in its own words, is a global, CEO-led organization of over "200 leading businesses working together to accelerate the transition to a sustainable world," including Unilever, Nestle, and Royal Dutch Shell.[62] Alongside the Gates-funded Alliance for a Green Revolution in Africa, they published a report behind the veil of the innocuous Food and Land Use Coalition, calling for "the right kind of investment in alternative proteins," alongside REDD+ and carbon pricing.[63] And indeed, the EAT-Lancet report makes many reasonable suggestions for dietary shifts. But within the Trojan Horse of erasing hunger through healthier food is a call for "a Half Earth strategy for biodiversity conservation," and as they add, "In-vitro meat production from cultured animal stem cells is being developed as an alternative to traditional meat." (Agrarian reform as an "alternative" to increase access to food in poor countries is not on the Lancet menu).[64]

Furthermore, the IPCC, merely mandated with supposedly non-normative technical evaluations, puts great stock in bioenergy with carbon capture and storage for many of its transition plans. Such plans are purportedly speculative. But they sketch out and in turn make reasonable or even thinkable certain futures.[65] This is so even though the IPCC always assiduously notes biofuels enter direct contradiction with people's need for food, the reason they have attracted the scorn of Third World peasant and popular movements.

The Energy Transitions Commission, an industry future-sketching outfit committed to green growth, draws up similar plans for forests, fields, and grasslands. On the ETC sit a pantheon of those most invested in fossil capitalism: Dominic Emery, the Chief of Staff of BP, John Holland-Kaye, the CEO of Heathrow Airport, and Chad Holliday, the Chairman of Royal Dutch Shell, the sectors most enthralled and in thrall to a permanent growth model, given gargantuan assets stored in the energetic source of future destructive growth. They do indeed gesture at how biofuels can compete with food, carbon drawdown, and biodiversity – a mandatory homage to the IPCC report, and a customary force field of caveats erected around any discussion of biofuels. But their program pointedly states, "Sustainable biofuels or synthetic fuels will need to scale up from today's trivial levels to play a major role in aviation and perhaps shipping," a slippage from cautionary notes to full-throttle embrace, inevitable if these programs are implemented. Likewise, the ETC national manifesto for the Australian leap to sustainability states, "Full decarbonisation for industries such as steel, cement and chemicals require the use of electrification, hydrogen, bioenergy and carbon capture and storage (CCS)."[66] The EU's energy transitions plan calls for massively increasing the biofuel mix in hard-to-decarbonize sectors like maritime and airborne transport.[67] The plan from the US Senate's Special Committee does too, alongside calling for afforestation.[68]

For India, where farmland for food has been dragooned into use for tropical export crops while Indian peasants starve, biomass is blocked out as a possible input into Indian metallurgy.[69] Using land for growing biofuels and planting trees for the express purpose of pulling in CO_2 from the atmosphere means not using such land for crops. Even under the most optimistic scenarios, shifting all of the world's hydrocarbon use to biofuels would cut savagely into agricultural production and water use. Furthermore, the slippage is not hypothetical. One quarter of the bio-based industry (BBI) projects are based on agricultural biomass, while a huge portion of the money from EU projects goes to agriculture-based biomass initiatives. Just 10 percent of biomass-based industries paid coordinators have anticipated positive results from biodiversity, and just a hair more than a quarter anticipate more sustainable management of planetary natural resources.[70]

The ETC also leans hard on carbon capture and storage: grabbing the CO_2 as carbon sources are incinerated or otherwise deployed and injecting it into huge caves underground. More ambitious yet is deploying capture and storage sky-vacuums *as* carbon scrubbers, pulling CO_2 directly from the atmosphere. In its big scope sketches, the ETC calls for capturing and storing "6 to 9.5 Gt of CO2 per year by 2050." That would require building 20 Mt of capacity monthly for 360 months. In total, carbon capture and storage (CCS) now captures just 33 Mt carbon annually.[71] A small hiccough: roll-out of CCS has stalled since it seems neither environmentally nor technologically prudent nor feasible. No less than BNP Paribas's investment counselor warns the great majority of CSS projects reinject the gas into oil wells in order to enhance recovery, a kind of Ouroboros of fossil capitalism.[72] But then, the ETC is helmed by the oil industry. Big money making more bad ideas.

CSS apostles run cover for the fossil fuel industry's desire to wring every last bit of profit from its petroleum holdings and infrastructure. Yet the broader buy-in relates to broader eagerness to maintain the physical-industrial infrastructure of modern capitalism, and to justify massive state investment in CCS with an eye towards preservation corporate valuations. And the emerging use for the physical land bases of Africa, Asia, and Latin America as carbon farms and for biofuels will allow for CO_2 offsets of whatever cannot be decarbonized and may allow for the continuation of a fuel-based modern and hierarchical order, at least for a few of the planet's people.

THE WORLD THEY WISH TO SEE

The World Economic Forum, the Allianz Global Investors, AXA, Enel, Goldman Sachs, HSBC, Japan's Government Pension Investment Fund (GPIF), Macquarie, and Michael Bloomberg, alongside Jeremy Rifkin, offer a "Great Transition" as a class project. Each call for an alliance between state finance, private capital/corporations, and communities, a Rorschach dot test in which people see what they wish. The word in fact obscures a central element of human community: communities are not homogenous social units, magmas of common interest, but contain people from different classes.

Goldman Sachs and Bloomberg are clearly part of a broader class project. Rifkin's call, a touch softer as befits a post-Occupy Wall Street moment, is for a "social market" economy, a riff on the German post-World War II experience. That historical moment rejected in equal measure a free play of market forces and command-and-control welfare of the Eastern bloc. The state would redistribute to protect social welfare only after the free to-and-fro of the market and competition had ensured adequate economic growth. Neither the CLIF nor the Rifkin GND are comprehensive. But their iron core is a form of hierarchy not so very different from the current one. Neither speaks of a turn to agroecology or says much on southern questions of development. Climate debt, the claim to humanity and justice of the southern nations from 1999 to 2009, is absent. If the olive branch of climate debt is noticeably absent from the northern technocratic-capitalist agenda for a green transition, the rifle is not. Rifkin calls for a mere 4 percent sliver of the US DoD spending to go to protecting the Earth. The CLIF does not directly mention military spending but does darkly gesture at southern questions of sovereignty when it notes that one primary concern for northern financiers is ensuring continued legal and fiduciary responsibility of southern governments for any form of green bonds or green infrastructural investment. Such commitments should be juxtaposed with the antipodean Breakthrough Institute's call for the national security sector to serve the greater good, or bad, of green transition via resource and human manpower mobilization.

There is no real progressive, let alone anti-imperialist, program even possible within a policy that insists on preserving the Pentagon and NATO, which will be necessary for frontier-preservation work when North-South inequalities swell amidst climate change as a "threat magnifier," and population movements and migration – a future of world-wandering climate refugees – are treated as people who need to be controlled and regulated rather than human beings who deserve dignity and development. More foreboding is silence around the raw material required for transition. Although all the initiatives speak of maximizing efficiency, none speak of reduction and redistribution, domestically and globally, of net energy distribution. Underlying this idea of a technological genie, who with a snap of the fingers can wish away hard choices about who gets what and who gets to live well and who does not get to live at all, is the idea of ecological modernization, the subject of the next chapter.

2

Change Without Change: Eco-Modernism

Can everything stay the same, while everything changes? Boeing 747s and Lockheed Martin F-16s burn biofuel from coconut husks and sugarcane in lieu of petroleum. Decommissioned factories recycle every molecule of metal. Internal combustion motors out, and lithium batteries in. Electric cars fill US highways. A national network of charging stations replaces ExxonMobil gas stations. Thorium nuclear reactors provide clean, meltdown-immune and limitless energy. Distributed solar panels and windmills carpet landscapes. Desalinization plants pock shorelines and water crises become a thing of the past. Greenhouses glitter in cold and dry landscapes. Carbon capture and storage machines vacuum CO_2 from the sky. Light doses of cloud brighteners and light-scattering particles cool off the Earth. As such technology develops, it gets cheaper. And because cheap and efficient energy was the basis for well-being in the northern industrialized countries, sharing such technological wealth provides for prosperity worldwide. Technological diffusion brings social progress within the framework of social market economies. This vision is the basis of purely technological "Great Transitions," swapping out the infrastructures of the fossil age for those of the Fourth Industrial Revolution.

The discourse of technological progress is the siren of industrial and post-industrial capitalism, distracting from political conflict over who gets what, while placing politics beyond the reach of the common person. The sweet but perilous song has two registers. The first lulls elements of a potential insurgency into the delusion that one need not decide in the here-and-now who gets what, since growth and progress will ensure that there will be enough for everyone eventually, if everyone, but especially those now without, just waits a bit. This reduces ecological politics based on what the environmental justice movement calls "distribution conflicts," to questions of waiting for technological advance.[1] The second register seems particularly attractive in northern countries but carries over into the

South, claiming that human-made tools can simply be repurposed by those in rebellion against the owners of the means of production. The issue isn't the knife itself, but who wields it and whose throat it cuts. Those who argue in an absolute way for technology's categorical social neutrality, especially from the left, forge one of the most dangerous, subtle, and effective instruments of ideological counterinsurgency: they accept the myth of progress and confuse opposition to the capitalist agenda. They suggest that every technology is exactly the same as the knife, the Kalashnikov repurposed by anti-colonial militia, or the printing press.[2] They argue that the new engines, machine tools, and social media sites are not welded onto the hands of the ruling class. And this is dangerous, because capitalists do not choose technology willy-nilly, but in order to maximize power, as has been shown time and again by critical historians of technology. Former MIT historian David Noble proved machine tools were designed to deskill workers and concentrate power.[3] Such deadly melees do not occur just on one shop floor or industry. Each technological edifice exists within a world-system of uneven accumulation. British textile production, access to cotton and wool, and markets for cloth and clothing, for example, were never merely a parochial British question. As the Indian economic historian Amiya Kumar Bagchi points out, cloth-making wove together social threads spanning the British slave trade, the underdevelopment of Africa, the deindustrialization of India, southern US cotton production, and British industrialization, into one piece of capitalist fabric.[4] Coal-fueled industrial take-off in Victorian Britain was one side of a coin whose genocidal reverse-face was Britain's late Victorian holocausts.[5]

MODERNIZATION THEORY GOES GREEN

If we follow these ideological growths to their bases, we see that they are not very novel at all. Technicism and its latest model, eco-modernism, are new branchlets from an old branch: ecological modernization theory, which split off the trunk of Cold War modernization theory. The ruling-class response to the climate crisis is not new but first germinated amidst a very different crisis: the threat of Communism, as the Soviet experience of industrialization in a single generation and Maoist agrarian revolution beguiled and bedazzled leaders and citizens of nations newly liberated from colonialism. The theory did not hide for a moment what it was about: *The Stages of Economic Growth*, by its demiurge Walt Rostow, carried the subtitle *A Non-Commu-*

nist Manifesto.[6] Rostow argued that societies would ascend through a series of stages on their way to modernity, roughly imagined as mirroring 1950s America: mass consumption, a mature middle class, suburbs sprawling around urban centers, a sedate and well-behaved political democracy. In this how-to manual for avoiding communism and achieving development, the benighted societies of the South needed a course of injections: Western technology, the book-learning which went alongside it, a proper Protestant entrepreneurial spirit, and a vibrant competitive urge. Markets, innovation, and foreign investment would merge, spiral, lead poor societies to sustained growth within the framework of the international trading system. These were also immunizations against Communism, which planners worried could infect southern countries weakened by debilitating lack of growth, and their leaders and peoples might look elsewhere for better ideas. But modernization theory was a fable. In reality, Western countries developed through catastrophic colonial and settler-colonial looting. And innovation was the collective gift of humanity to future generations, and arose throughout the world.[7] Although modern science arose in Europe, human innovation itself has never been the monopoly of the West.[8] The power that fueled the "modernization" engine relied on primitive accumulation of atmospheric space.[9] Western wealth was largely based on slavery and stolen land used to grow cotton and sugar, and colonial extraction from the agricultures of India and Indonesia.[10]

Latin America, the chunk of the Third World where direct colonial control had been thrown off earliest, was the nursery for the earliest ideological counterattacks against modernization theory: structuralism, and dependency and world-systems theories, crafted by the great southern economists and historians: Brazil's Celso Furtado and Vania Bambirra, joined soon after by Egypt's Samir Amin and Guyana's Walter Rodney.[11] If the dependency theorists mounted one attack against the myth of modernization from the perspective of the South, the ecological movement mounted another, from the perspective of the future and the perspective of those poisoned by pell-mell spilling of toxins. The hydra of the environmentalism of the poor grew many heads: from Black anti-landfill protests in North Carolina to Cesar Chavez's farmworkers organizing against pesticides in California. These revolts of the 1960s–1980s rose against the world-eater of nuclear apocalypse, rising awareness of the pesticide plague, and swelling understanding of industrial agriculture's evisceration of soils and their fertility.[12] Northern spokespeople included the biologist-philosophers Barry Commoner and Rachel Carson, who called for a humane relationship with the natural world. Countries of the South produced their own spokespeople: Tunisia's Azzam Mahjoub and Slaheddine el-Amami, India's Vandana Shiva, Mexico's Efraín Hernández Xolocotzi and Victor Toledo.[13] And others who from an early point developed an anti-colonial ecological discourse with care and caution and worry about how northern ecological chatter was burbling with Malthusianism, eager to turn environmental concerns against rights to development.[14]

Then came the counterattack. Ecological modernization theory began to grow into a dangerous nightshade in the 1990s. It argued economic growth and ecological health did not need to be at odds. Amidst the "End of History" and the electric triumph of liberal capitalism over the Soviet Union and what it represented, its ideological spinners and spanners tried to rip the thread of ecological thought from its humanist and Marxist fabric and reweave it into a natty green garment for a shopworn and dirtied capitalism. Ecological modernization theory became the basis for a synthetic politics. Not that old dream of swirling the green of radical ecology into the red of socialism, but a cheap fluorescent veneer to cover ideas of growth, modernization, capitalism, democracy, and development. Arthur Mol, an environmental sociologist and chief evangelist of this approach, argues that humans can make "an environmentally sound society" without including other social qualms, such as "the scale of production, the capitalist mode of production, workers' influence, equal allocation of economic goods," or any gender criterion.[15] The theory of ecological modernization came from speaking "the proper language" of the "dominant institutions," in the words of Maarten Hajer, a Dutch political scientist.[16]

The proper language is "the language of business." Such theories were long on hypotheses and short on evidence. They appear and reappear with mechanical regularity, in new forms and proposing new gewgaws and technofixes, and refusing to engage with the mountain of facts which disprove their claims. As the environmental sociologist and editor of *Monthly Review* John Bellamy Foster notes, the academic (and certainly popular) prominence of such social science fiction "can be viewed as an indirect acknowledgement of the structure of power itself."[17]

ECO-MODERNISM

Ecological modernization theory was and is intended to discredit, encircle, and counterattack against the advance of critical and humane ecological scholarship. Eco-modernism, garbed in communist crimson or nakedly capitalist, has lately ventured outside academic cloisters and advanced in more popular arenas, including those identified with the left, building a theoretical and intellectual bridge between capitalism and purportedly socialist schemes.

Eco-modernism had its early popular sketches in self-described environmental writers Ted Nordhaus and Michael Shellenberger's pamphlet, *The Death of Environmentalism.*[18] The polemic presaged contemporary sneers at the old ecological movement that cleaved social concerns from a nebulous nature (in the process evaporating from consciousness the more widespread if less visible environmentalism of the poor). Ideas need not be directly vanquished if they can be murdered through ignorance. Their book was an assassination attempt against traditional environmentalism, an early gambit to reintroduce Malthusian thought in its place with their worry that "overpopulation" was driving the ecological crisis, and "economic development" ought to be fused with environmental concerns. Here was the intellectual vanguard for the "green jobs" jamboree that became the stock-in-trade of green capitalists and social democrats alike in the late 2000s and 2010s.

Such talk is magnetic for a forward-looking investor class interested in preserving the financial architecture and hierarchical soullessness of capitalism while discarding the blunderingly ecocidal elements of endless accumulation. They soon built a comfortable home at the Breakthrough Institute, where they sit alongside other Breakthrough in-house columnists, such as the Hyatt heiress Rachel Pritzker, who is on the board of the Energy Growth Hub, alongside Francis Fukuyama of end-of-history infamy, and where Toss Moss, a former US Deputy Assistant Secretary of State, is the Executive Director. They are also authors of the *Eco-Modernist Manifesto*, which grabbed wide media attention in 2015.[19]

Its foundation stone is decoupling, which argues that human development, a slippery stand-in for growth, can be sheared from environmental impacts. The authors argue that humanity must lighten such impacts "to make more room for nature." They do not think that "human societies must harmonize with nature to avoid economic and ecological collapse."[20] They consider this a strange romance. Indeed, rather than the immensely complex, artisanal-humane and eco-socialist goal of living with nature, they argue: "Natural systems will not…be protected or enhanced by the expansion of humankind's dependence on them for sustenance and well-being." Humanity should dwell in an austere and antiseptic techno-sphere, dematerialized and drawing on phantom technologies that lessen human impact on the biosphere to the lowest possible level. We are already partway to a featherlight impact of farming, energy extraction, forestry, and settlement, as humans are "less reliant upon the many ecosystems that once provided their only sustenance," although to this day the history of development has "deeply damaged" non-human nature.[21] (In this, the eco-modernists have a far less sophisticated ecological analysis than the World Economic Forum or other co-thinking architects of the next prison.)

The platform finds space for every part of the current social order in the world to come: entrepreneurial capitalists, the state, markets, and civil society. They propose a suite of technologies, none of which exist or are scalable, to replace each pillar of human life. For energy, a closed uranium or thorium fuel cycle, or from hydrogen-deuterium fusion or to capture the essentially limitless solar radiation which falls onto the Earth. For settlement, putting humans into cities, which "symbolize the decoupling of humanity from nature," saving land for the rest of life.[22]

This peculiarly Western radical separation of humanity from the rest of nature finds its accomplice in the notion of "land-sparing agriculture," perfectly exemplified in the dizzyingly productive, dazzlingly labor-light industrial agriculture of the US. As farming has become "more land and labor efficient," rural people have left the countryside for the metropolis. As yields increase, fields retreat and forests advance. Eighty percent of New England is forested today, as compared to 50 percent at the end of the nineteenth century. Such trends, they argue, are the future: human impact on the environment can peak and then decline in the twenty-first century.

While they nod at the notion that resource extraction is still occurring, just over land and sea, they argue that such remaining footprints can be lightened and erased through developing more efficient mechanisms for procuring needed resources: urbanization and nuclear power, desalinization and aquaculture, and agricultural intensification.[23] At the core of the manifesto is cheap energy, which makes for cheaper recycling and cheaper capital-intensive and energy-intensive inputs on farms. Finally, these demands are the ore with which a steel sacred calf is cast and at which all subsequent eco-modernist literature worships: meaningful "climate mitigation is fundamentally a technological challenge." In other words, the only items on the menu are their tech, or a return to a pre-modern atavism, which they bizarrely suture to the very emblem of Cold War dwelling patterns: "suburbanization," counterintuitively melded with "low-yield farming, and many forms of renewable energy production," which sprawl out on land, which could be better used as enclosures or colonial fortresses for a pristine nature.[24]

Much like the various Fortune 500 planners, they want to densify population to preserve "wild" nature. Following their density fetish, they dismiss distributed solar power, a relatively democratic and humane pathway to planetary access to electricity, as mauling nature. Eco-modernism washes itself in green and becomes a backdoor through which nuclear fission becomes not one amongst many technological choices but the only possible technology – a green which glows in the dark. Until then, dams and (imaginary and oil-industry-linked) carbon capture and storage tech attached to power plants are a serviceable backup. They conclude by defending modernization, distinct from capitalism and "corporate power," and rejoin it when they defend the leading role of entrepreneurial capitalism. Inverting the extensive actual history of modernization, wherein capitalism has led to steadily increasing material thorough-put, appropriation of primary production, and increase on nearly every index of impact on the non-human world, they assert that modernization has allowed human societies to meet "human needs with fewer resource inputs and less environmental impact."[25]

As with other new-fangled technocratic musings, these shiny trinkets are not very new at all. The dream of radical separation from nature is an apartheid concept tracing back to early capitalism with its mechanical worldview and extreme violence which disassembled pre-capitalist organic unities.[26] Such colonial walls between human and nature took form in the US at least as far back as the creation of Yellowstone National Park on stolen Indigenous land.[27] The manifesto, then, offers a fluently mendacious millennial jargon for modernization theory.

These specters reappear in subsequent documents, with their emphasis on technological change as a way to sidestep basic distributional problems, or hard puzzles over how to structure the world in which we live so as to use the means of production to make people's lives better in the here-and-now. Perhaps the most emblematic example of this type of thought is Aaron Bastani's *Full Automatic Luxury Communism.*

ACCELERATIONISM MOVES LEFT

Bastani's manifesto is at core a communist version of the eco-modernist manifesto. Instead of redistributing ownership and use of what we currently have, we should instead grow wildly and then redistribute that. This is a call for a world in which "abiding scarcity," or even having to sometimes work hard in order to produce and reproduce human civilization, does not blight "post-capitalism." Rather than reworking human societies to function better on less industrially sourced energy, we should rethink how we get critical inputs like lithium and metals so that we can build billions of batteries and solar panels and turbines, and use them to fuel a world where we do not have to work very hard at all. How to conjure up this world? "We'll mine the sky." He is highly optimistic about the prospects for decreasing costs and increasing capacities for space exploration. Using such cheaply available tech, we can forage amongst the asteroid belt for iron, platinum, and nickel. "According to one estimate," he notes, "the mineral wealth of NEAs – if equally divided among every person on Earth, would add up to more than $100 billion each."[28] Tellingly, most of Bastani's copy material is from often government-subsidized venture capital firms. Is sifting through the solar system for platinum, cobalt, and iron really on the cards? Let's check with NASA. A study provided on its website concludes:

> There is no economically viable scenario we could identify that depends solely upon returning asteroid resources to LEO or the surface of the Earth. To be economically feasible, asteroid mining will depend predominantly upon customers in-space who are part of the space industrial economy and infrastructure.[29]

The report adds, "It is possible that secondary markets on Earth may provide 'icing on the cake' for platinum-group metals." As for other resources, "Rare Earth Elements [REE] are not an Economic Option...Although the initial... proposal included REEs as a potential option, there is not economically advantageous basis for returning REEs from Space to the Earth."[30] So much for asteroid mining.

For agriculture, Bastani somewhat incredibly resurrects the Green Revolution as an exemplar of modern agriculture. Using Green Revolution technologies, he argues, we could make "food so abundant that – as with energy, labour and resources – it would become virtually free," with value emerging from "informational content" rather than "land or human effort." Through using capital-intensive and energy-intensive inputs like massive applications of fertilizer, pesticides, and herbicides, and using semi-dwarf wheats that are highly responsive to those inputs as well as added irrigation, "wheat yields have tripled...possibly saving a billion lives." And he goes on to ask, "What if that Green Revolution...was only the beginning? What if... we had only begun to understand how our mastery of nature could confer almost limitless abundance? ... [W]hy should hunger exist at all?"[31]

What Bastani means by all of this is hard to say. We now know with surety from the work of scholars like Harry Cleaver, Raj Patel, and Divya Sharma that the Green Revolution was not designed to end hunger, nor did it end hunger.[32] Rather, it was intended as one resolution to the resurgent Maoism haunting the Asian continent. In the words of William Gaud, then the head of the US Agency for International Development, a "Green Revolution" in lieu of a "Red Revolution."[33] It was, furthermore, a strategy of "betting on the strong": diverting state resources to the medium and wealthier farmers capable of more easily using the new technologies. Wheat yields did increase, but so did landlessness. And if you don't have a wheat plant, increasing wheat yields will not make you less hungry. Decreasing rural labor when there are not urban jobs will, for that matter, make people hungrier, not less hungry. Furthermore, nutrition is not just about quantity of calories, but also their quality. As rice and wheat replaced millet and other "coarse cereals" there was a "deficiency of micronutrients" in Indian diets.[34] What it means to save a billion lives is also not clear. As Richa Kumar notes, "There is inadequate evidence to support the claim that India was food-insecure in the 1960s. Moreover, evidence suggests that India's food and nutritional insecurities today are the aftermath of the green revolution strategy promoted since the 1960s."[35] Other paths were then and remain open: a more labor-intensive smallholder-centered path based on the resurrection not of highfalutin myths about Green Revolution productivity, but of the more traditional landraces of India, and other Third World coarse grains and rice varieties.[36]

Finally, consider a third artifact of Full Automatic Luxury Communism: lab-meat. Bastani argues that "animals remain energy intensive and inefficient at converting solar energy to food," comparing Bangladeshis eating a vegan diet to Americans eating 270 pounds of meat.[37] Yet, not all is well in Bangladesh. In its rural regions, according to the Food and Agriculture Organization, "Rates of malnutrition...are among the highest in the world.

More than 54% of preschool-age children, equivalent to more than 9.5 million children, are stunted, 56% are underweight and more than 17% are wasted," with pronounced macronutrient deficiencies.[38]

More to the point, animals do not exactly convert solar energy to food. They eat plants. Different plants grow, however, in different places, and grow differently on different kinds of land, which is why the category "land," unproblematically deployed by Bastani, does not help us understand what can or should be grown on which types of land. He cites a Cornell study which claims that one-third of the planet's surface is used for livestock, and 302 million hectares are used for livestock in the US. Meanwhile, just "13 million hectares were allocated to vegetables, rice, fruit, potatoes and beans. Such a huge gap shows that animal products are a highly inefficient way of using finite resources to produce food."[39] This is logic for the illogical. The actual Cornell study states that of those 302 million hectares "272 million hectares [are] in pasture and about 30 million hectares [are] for cultivated feed grains," and recommends "cattlemen switch...to grass-fed production systems."[40] In 2016, 407 million US hectares went to agriculture and pasturing.[41] So, about one-fourth of US cropland goes to feeding animals. That is because cows mostly eat grass, and land used for growing grass is not good for growing other things that humans like to eat. In short, Bastani misrepresents the facts and the argument.

Building on an intellectual foundation made of quicksand, Bastani argues that "it is clear that the world needs to eat far less meat than it does. Preferably, we would completely eliminate it from our diets."[42] He is not talking merely about US or EU meat-gobbling, but about the entire planet: "the world." What would we replace it with? Cultured meat. Yet cultured meat needs energy to grow, whether from oil, gas, or solar panels. At least theoretically, animals are very good at converting photosynthetically derived energy from the sun into a form usable by humans. As another study notes, "in vitro biomass cultivation could require smaller quantities of agricultural inputs and land than livestock; however, those benefits could come at the expense of more intensive energy use as biological functions such as digestion and nutrient circulation are replaced by industrial equivalents."[43] In other words, in an already overindustrialized planet, Bastani wants to replace further biological-organic with industrial processes. Furthermore, he wants to do this on a planetary basis, which would require massive displacement or intervention in the lives of pastoralists and smallholders who live off sustainably grazed meat across the Third World.

The through line of these three interventions is the assumption of categorical technological neutrality: the issue is not the knife, but what it cuts, alongside any concern for those lives the technology has cut or will cut apart: rural people in the Third World. Whatever the intent of such social science fiction, the result is to splash red paint on capitalist technology, and in the case of the Green Revolution, to actually carry out radical historiographical revisionism and introduce reactionary thought into the left. The other outcome is to disorient and disorganize opposition to the actually existing agenda, outlined in Chapter 1, of the big foundations and their backers, who want to strip poor people of the right to eat any meat at all.

DEBUNKING DECOUPLING

At eco-modernism's core is the claim that technologies are categorically socially neutral. The foundation stone of these arguments is that it is possible to decouple at speed, to gradually dematerialize the global economy, so that it can eventually run on closed loops, and that green growth is possible. Giorgos Kallis, a Greek ecological economist based in Barcelona, and Jason Hickel, an anthropologist at London's Goldsmiths, take this argument head-on, asking: "Is Green Growth Possible?"

It is true that a small handful of northern economies are slowly breaking off GDP growth from CO_2 growth. But this truth is also the fruit of statistical arbitrage and a form of carbon-chicanery: how international monitoring institutions count carbon emissions.[44] They seldom focus on where CO_2-laden commodities are consumed, but on where they are produced. Thus, in theory, China, Vietnam, Bangladesh, and the Philippines could be the site of nearly every polluting bit of industrial production currently occurring in Finland and France, while the entire WTO trading architecture and imperialist political engineering, and unequal exchange through which value and wealth concentrate in the North, remain intact. Chunks of high-end programming and marketing, sophisticated work on robotics, and patents could remain in the North. The result would "decouple" GDP from CO_2 in the affluent core. Such a sleight of hand is based on imperial trading architecture and carbon accounting procedures. In fact, the trend towards decreasing impact of northern development on northern environments has occurred alongside increasing impact of northern development on the southern environments. Four to eight percent of air pollution in China in 2006, for example, was linked to exports of products to the US. Global particulate matter is increasingly thrown into local atmospheres from local production, whereas the goods produced are consumed elsewhere.[45] Contrary to over-hyped claims about the productivity of northern agriculture, food is a huge driver of this breakdown, as the clean and good and low-impact food eaten in some places has enormous impacts elsewhere: in 2007, 50 percent of the EU total consumption of cropland, grazing land, and forest land took place in other countries. EU and USA patterns of consumption lead to nitrogen pollution worldwide in water tables.[46]

The lightened rural impact of the US population goes hand-in-hand with mounting impact on the potential croplands of the former Third World. Nor is dematerialization of production possible: windmills and solar panels cannot be made without industrial production processes, and windmills have negative effects, which means there should be popular consultation as to where they are sited.[47]

The core has not at all achieved absolute decoupling if one takes into account how CO_2 dumping industries have been relocated to peripheries and semi-peripheries. Relocating smokestacks and smelters to the South has meant factories move, but value flows are the same. North–South income gaps have steadily widened, according to the work of world-systems scholars Giovanni Arrighi and Beverley Silver of the Johns Hopkins University.[48] Similarly, Kallis and Hickel show every single model for continued growth alongside reductions in CO_2 emissions relies on technologies that do not yet exist. Like the joke of how an economist proposes to open a steel tuna can on a desert island by assuming he has a can-opener, they assume incredibly fast reductions in CO_2 emissions in the wealthy core, paired with bio-energetic carbon capture and storage, bio-energy carbon capture and storage (BECSS), a negative emission technology based on tremendous tree monocrops, which are burned for energy. Imagine where BECSS will be sited in a world where national sovereignty and the national question are dismissed as irrelevant? (As the IPCC elsewhere warns, if they are deployed at the multi-gigaton scale, they could lead to "adverse side effects for adaptation, desertification, land degradation and food security".[49]) These scenarios allow for "overshoot": exceeding the carbon budget. Of scenarios which exclude BECSS and target 2°C of warming, global emissions need to hit net zero by 2075. For 1.5°C they need to hit net zero by 2050. These models assume a full shift to renewables, rapid soil regeneration and afforestation, and industrial processes for metallurgy and cement which do not require emissions. It is possible that such technology can be installed fast enough to fully replace the existing emissions-emitting industrial plant. To do so while maintaining growth will be difficult. To do so while sharing justly electricity resources with the Third World is impossible, as I argue in the next chapter.

CONCLUSION: LEFT ECO-MODERNISM AND TECHNOLOGY

So, what is at stake amidst this flurry of confusion? Ideas do not come from nowhere. They emerge from certain contexts. Some flourish and others wither depending on how well they are able to germinate and send their roots into distinct social contexts. And they certainly do not spread by virtue of being correct.

Much eco-modernist theology runs against the grain of everything we know about environmental and economic history, but it gains more and more in popularity. And that includes within the First World Left. That should not shock us. First, ecological modernization theory is counterinsurgency in the realm of thought, meant to contain, dominate, and subvert autonomous thinking about the common life, or communism, by forcing the oppressed to fight on the terrain of the oppressor. Modernization theory accepted the meta-notion that the good life should be democratically shared, but brought with it the coercive falsehood that the good life in the North had not been based on the bad life in the South – capitalist development and colonial de-development as two sides of one historical coin. Ecological modernization theory accepted capitalism was damaging the ecology, but argued such damage could be remedied and was being remedied by the tendencies of the system. Underpinning the older theory and the new were two false claims: one, that capitalism is not inherently polarizing and exclusionary; two, the technologies accompanying specific paths of capitalist development were socially innocent, rather than deliberate weapons in the class war.[50]

Although these aggressions rest on specific configurations of technology, they are not the same thing as technology, nor do they lead to the political syllogism that because a technology is used to attack the people, the people should reject technology. Indeed, no one this side of John Zerzan or Derek Jensen rejects "technology." This makes it maddening and muddying that many in the green social democrat camp contribute to rather than clarify such confusion while trying to find a third way between eco-modernists and their critics, as they argue that anyone, anywhere, is against technology. As the historian of technology Noble points out, such frameworks do not do a great deal to help us approach the issues at stake: "Technology as such does not exist. Technology exists only in the particular, as particular pieces of equipment in particular settings."[51] What people can (and should) reject are specific technologies. Technologies are infusions of knowledge into material goods, whether ores, wood, cottons, jute, or jasmine, which have prices, which are the effects of global power relationships.[52] Technology is not abstract but concrete. Under capitalism, it is very broadly introduced from above. New technologies are not manna. Management adopts them to control workers and peoples, and to maximize the ceaseless accumulation of capital. The monopoly of the concept of technology by the ruling class and its clergy and its deployment as an abstract justification not merely for the imposition of new technologies but a call for acceleration of their introduction is in form and substance an ideological surrender – one which more cautious contemporary democratic socialists also accept, or certainly refrain from contesting. It is part and parcel of the myth of progress, of things perpetually getting better, which lies at the root of every modernist communique, polemic, and pamphlet, and is a weapon meant to disarm anti-systemic movements of their most potent weapons. To assert that anyone in particular is arguing about technology in the abstract is to sharpen the edge of a weapon designed to disembowel resistance to capitalism.

It is those claiming that technologies are socially innocent who have the burden of proof, for theories of imperialism and environmentally uneven exchange show that they are not socially innocent. Furthermore, more important is how eco-modernist theories, whether of the right or the left, are blind to the needs of the environmentalists of the poor in the peripheries of the world-system, and have neither time for nor interest in the social and ecological demands from the world's peasant movements. Indeed, politically, eco-modernism and red eco-modernism have been silent on the great achievements of the environmental justice movement from the 1990s to 2015: putting climate debt, ecological debt, and reparations onto the world agenda; enshrining in the highest-level climate summits the idea of common but differentiated responsibility, or that each nation must do something about stopping CO_2, but that such responsibilities might differ in accord with historical responsibilities.

CO_2 emissions can be stopped. The technology exists to replace portions of unsustainable energy use with sustainable energy use, and to some extent

3

Energy Use, Degrowth, and the Green New Deal

Societies run on energy. Older societies were powered by the sun. Plants used photosynthesis to convert solar energy into living matter: wood for fuel, feed for animals, and seeds and fruits for people. The widespread use of coal and oil ushered in a different way of organizing human society, based on massively increased growth in energy use, and growth in the human-built paraphernalia that channeled hydrocarbon-based energy. Widespread human well-being in the northern countries went alongside continuous growth in GDP and private property. Predictably, if far from logically, there has emerged an idea that healing the ecology and humankind's place within it has to be based on continued growth and, it goes usually without saying, private ownership of the means of production, to keep humans healthy, happy, and prosperous.

Against this background, it is no surprise that social democratic GNDs have been about greening growth: green job guarantees, green infrastructure, green tech, green cities, green agriculture, all against a shimmering but elusive background of more. Amidst and against such a notion, a rising European-American degrowth discourse suggests what we need is less. Growth ideology justifies austerity, neoliberalism, and consumerism. It is good to rid ourselves of such a delusion since it dispels the soothing cliché of melioristic tides lifting all boats to show which boats are made from the shattered timbers of other ships.

But actual elite planning is currently ambivalent on growth, given the Malthusian enthusiasm of Western foundations, their fretting that some dark populations are growing a bit too much, and finally their actual practice in places like the Arab regions: de-development, a phenomenon first identified in colonized Palestine, part of a longer history of colonial apocalypse.[1]

One can imagine three short-term ways of orienting progressive practice, each vying for dominance in the current moment: one, full renewable replacement of current energy use, alongside increased southern energy use, and ongoing capitalist property structures – the left-liberal solution. Two, lower energy use through retrofitting core countries' infrastructures, substantial domestic redistribution to go back to 1950s-era levels of (in)equality, alongside full replacement of the energy infrastructure, and an ambiguous call for grants to help southern countries transition – the green social democratic and being-nice-to-the-South solution. And three, considerably lower energy use in the core alongside decommodified social infrastructure, guaranteed well-being, and massive technology grants to the Third World. A form of (degrowth?) eco-communism. The paths are stylized. They are more like a million rays, each a hair's-breadth different from one another. The more redistribution there is in the core, the closer path one looks to path two. The larger the grants are, the more path two looks like path three.

DEGROWTH

Degrowth is a political-ecological call for sufficiency. It is based on understanding that the weight of capitalist technology on the non-human world is more than that world can bear. The intellectuals and activists clustered around degrowth, especially its Iberian promoters, have been massively successful in throwing a series of ideological monkey wrenches in the secular faith in growth, an ideological glue which binds together people in the Western capitalist pseudo-welfare states. The degrowth intellectuals' perspective carries valuable water in northern industrial capitalist states which must reduce overall material and energy use, and where populations increasingly value the environment over economic growth.[2] However, at times, at least, degrowth has also been an ecumenical camp, with some degrowthers focused more on growth rather than accumulation and social hierarchy, while imperialism has not always seriously figured in the degrowth position.[3] Furthermore, eco-socialist thought pre-dated degrowth, and movements of the environmentalism of the poor pre-dated the formulation of eco-socialism.[4] So, I prefer to frame the question a bit differently.[5] Some sectors, such as agroecological food production, public transport, primary healthcare, and renewable energy, need to grow incredibly fast. They must do so while remaining decommodified. Others must disappear: the military, non-renewable energy production, chemical fertilizers.

ENERGY USE AND WHY WE MUST DOWNSHIFT

Energy use is the meta-sector, threading through so many other sectors. Shifting to different patterns of energy use and different sources of energy conditions shifts in every other social sector from consumption to production – corporations to cooperatives, households to individuals. A just anti-colonial socialist resolution to the climate question is not possible without massively and immediately shifting the power systems of industrial civilization.

Shifting technological or social systems means shifting social power. From highways to the automobile industry to the current farming system, an entire world has been built in the core countries on economically "cheap," physically dense, and easily storable forms of power. Social scientists call this the "fossil energy regime." As the energy researcher Simon Pirani points out, "technological systems," like mass automobile use, plastics and petrochemicals, "have evolved in the way they have because of the social and political systems in which they are embedded." We in the core states live amidst gargantuan systems which are inseparable from the use of the dead, fossilized photosynthesis of eons past, carbonized into black blocks of coal and oily black goo.[6]

What does that mean for systems change, climate change, and energy-systems change? We have to downshift current US levels of energy consumption for a globally just GND. Recall here that the US uses around 12,000 kilowatt hours per person per year, Japan uses 7,150, and France uses 4,928. Meanwhile, Iran uses 3,072, Nicaragua 571, Sudan 268, and Yemen 91. Such differences are not just.

Different paths for CO_2 reduction concern how much harm will be inflicted on the planet and its poorest inhabitants. They pose two explicit questions to people in the North: one, how much are we willing to change our lives to ensure justice for all of humanity? And two, are we willing to change our lives and attack a system of accumulation which will without question lead to the hammer of climate change clanging down hardest on the world's poorest? They pose a third, implicit question: are we capable of seeing the latent fascism even in halcyon social democratic models which leave open the possibility of intervention, exploitation, or extirpation of the Third World, and by leaving capitalism alive there, give it safe haven for attacking middle classes in the core itself – the phenomenon often called neoliberalism?

MYTHS OF THE CARBON BUDGET

One way climate debates are made legible across the political spectrum is through the idea of the carbon budget, which scientists use to explain how much CO_2 we can emit to stay below a pre-determined "crisis point" – in quotation marks because, for most of humanity, the crisis long antedates climate change. One thing about the carbon budget is that the amount of CO_2 humans can safely emit keeps decreasing. First, because capitalism keeps emitting CO_2. Second, because scientists' modeling and the way they report their results each have conservative biases.[7] Third, because scientists learn more and more about the consequences of each degree of warming, and accordingly reduce the "safe" increase in world average temperatures.

Each of these points highlights how much politics are embedded in the "objective" science of carbon budgeting and temperature limits. Consider that just in the past decade, scientists have switched from a general "consensus" of the need to target 2°C to the need to target 1.5°C, because the social consequences of the higher number versus the smaller one now impel natural scientists and the ultra-conservative IPCC process to understand the logic of at least speaking in the public arena about the need to aim at 1.5°C.[8] Others, writing in the flagship journal *Nature*, suggest even 1.5°C "may cause a substantial thaw of continuous permafrost as far north as 60°N."[9] Such a fact does not prescribe any specific political strategy, but it does clarify the consequences of distinct political strategies and the emissions reductions with which they are paired.

To show why this is the case, we should look more closely at the carbon budget. Take the most recent estimates, or the IPCC SR15. For a 66 percent chance of avoiding a 1.5°C in warming, they allot 420 Gt of CO_2. That is around ten years of current emissions, which has given us the media-ready number of 2030 to reduce emissions to zero. For a 50 percent chance of going over 1.5°C, they have a quota of 580 Gt of CO_2 or 14 years of current emissions. However, as the report states, "uncertainties in the climate response" to CO_2 and non-CO_2 "emissions contribute" ±400 Gt CO_2. Very simply, there is one-in-three chance that if we burn that much carbon, we will go over, maybe well over, 1.5°C of warming.[10]

Much of left-liberal climate talk is based on administering rather than eliminating capitalism, and as a result is built on a seldom acknowledged foundation of assumptions regarding the global distribution of wealth and consumption, and the institutions with which it is tied, in terms of why emissions are produced and their consequences, which are intimately related to which lives matter and which lives do not. For that reason, it often does not dig too deeply into these scenarios or the capitalist context in which they unfold, and in which, even focusing merely on climate harm, will continue to disproportionately target the periphery's poor.

Take an impossible scenario: all emissions stop today. Even if that were to happen, people in the South would continue living in the climate of injustice forged by historical CO_2 emissions. For some sense of those killed or made homeless by climate-related disasters, note that from 1980 to 2002, the total was 300,964 killed in Ethiopia, 168,584 in Bangladesh, 150,362 in Sudan, and 101,473 in Mozambique. During that period, 62,553,000 were made homeless in Bangladesh, as were 8,679,282 in Pakistan and 7,823,102 in the Philippines.[11] As Oxfam reports, on average, over 20 million people a year were internally displaced by extreme weather disasters over the last ten years. Small island nations like Cuba, Dominica, and Tuvalu, archipelagos like the Philippines, and immiserated states like Somalia are amongst the ten countries where people are likeliest to be afflicted by climate-caused disasters like cyclones and floods. Not one of them is in the top 95 emitters in the world.[12] As filmmaker Rehad Desai notes,

> Ninety per cent of Mozambique's second largest city was destroyed, the construction of 300,000 homes is now required and yet to begin, the price tag $600 million, roads and public utilities $700 million. The death toll of 1000 only took into account those killed by the immediate impact of the cyclones. Yet to be counted are those that perished following the spread of hunger and disease. The unpalatable likelihood is that Mozambique will never fully recover from the destruction created by cyclones Idai and Kenneth and that furthermore, the country will be hit again in the near future.[13]

Elsewhere, there are plans to actually relocate Jakarta, one of the biggest cities in the world. Nor is such disaster limited to the periphery of the planet. Hurricanes Katrina and Sandy hurt poor and especially Black people in the US and were almost certainly made more violent by climate change. The 2019 atmospheric sequence which led to the historic floods in the Midwest, from frozen ground to bomb cyclones, were also probably the product of planetary climactic shifts.[14] The seemingly apolitical and technocratic construction of the carbon budget papers over that every decision made by capitalists in the North to continue to emit or to only slowly reduce emissions is a decision which destroys entire cities in southern Africa or sweeping swathes of rural America. Maps of which regions will be most devastated by climate change overlap with poverty, and with relative concentrations of the Black population.[15]

Furthermore, decisions about carbon budgets tend to ignore that capitalism is a climate of permanent emergency and structural violence for the planetary majority. Probabilistic administrative and technocratic decisions reflect the notion that the emergency is over there, on the far shore, over the horizon, and should not be included in domestic political calculus. The struggle over climate debt, which is the topic of the final chapter, was a partially successful attempt from the world's periphery to introduce reparations within climate treaties. These were ways to acknowledge and try to repair the historical and contemporary damages inflicted by capitalist accumulation. And these discussions did not sidestep the state of emergency in which the poor periphery lives, not as a result of the climate emissions of today, but because of the emissions of yesterday, the neo-colonialism of the years before, and the colonialism of the years before that.

Furthermore, the structural violence embedded in the sterile arithmetic of carbon budgets reflects an equation that avoids quantifying the interests and aspirations of all relevant parties. There is an intergenerational component of injustice in carbon budget mathematics, which do not adequately weigh and balance questions of happiness and sacrifice in the present and in the future. Political agitation for a just transition would include explaining to people that some kinds of things must be given up in order to secure the world for their grandchildren. Such explanations would expose scarecrow rhetoric of "sacrifice" and "austerity." Even under capitalism, every human understands the desire and urge to leave wealth to their children in order to make their lives easier and more comfortable. Shifting the entire productive structure takes that common sense of intergenerational care and concern and enshrines it as the ethic of conscious social planning.

Another big problem with carbon budgets: they are probabilistic. They are not certainties. We cannot be sure that a given quantity of carbon will do what the models claim it will do. For that reason, human safety would require politically difficult decisions, which are generally removed from the discussion in advance. Suppose a given quantity of carbon has a 70 percent chance of holding warming to 1.5°C, it might require a far smaller quantity of carbon to have an 85 percent chance of holding warming below

1.5°C. According to existing models, there is no amount of carbon which would give a 100 percent chance of holding warming below 1.5°C. These are not hypotheticals: according to the IPCC SR15 report, the most optimistic models give a carbon budget of 420 $GtCO_2$ for a 66 percent chance of avoiding 1.5°C, 580 for 50 percent, and 840 for 33 percent. They found that "No pathways were available that achieve a greater than 66% probability of limiting warming below 1.5°C during the entire 21st century based on the MAGICC model projections."[16] And to get a sense of the ecological damage which the models bake in, we can see that to have a 66 percent chance of keeping warming below 1.3°C, we would have 80 $GtCO_2$, or two years of emissions. It is understandable that such numbers are difficult to metabolize politically, but it is important to do so from the perspective of those who will suffer from further warming.

THE PRECAUTIONARY PRINCIPLE AS RESPECT FOR WORKING-CLASS AND FUTURE LIFE

These numbers imply a radical scale of change. First, consider not the 66 percent, but the 33 percent, which ought also to be avoided, and furthermore, the nested probabilities in that 33 percent. Some argue there is, for example, a chance we have already entered the stage of "runaway" climate loops, where based on the current amount of atmospheric CO_2, the average world temperature will increase 2°, 3°, or 4°C.[17] Human response to climate change ought to proceed on the precautionary principle. A group of scientists explain the precautionary principle this way: "The precautionary principle, by calling for preventive action even when there is uncertainty, by placing the onus on those who create the hazard, and by emphasizing alternatives and democracy."[18] If one considers that probabilistic estimates suggest a 5 percent chance of runaway warming given carbon budgets for 1.5°C, the conclusion is to put all human energy to work in a just transition, and to move as fast as possible.

Debates about possibility will necessarily be political and cannot be anything else: people will not move politically on a large scale to reduce emissions so radically if it means real austerity and a pre-industrial and pre-modern lifestyle, without ovens, without any metals, without any industrial forms of transportation at all. Determining what is "as fast as possible" is a political-technical decision, and one that requires political and social power in order to act upon. There is no objective reason why large-scale private property regimes should continue, for example, or that popular classes should be unable to socially plan the present and the future. Those are political decisions. Accepting the continuation of monopoly private property is a political decision. Choosing to struggle against it is a political decision. And choosing to partially reform it or channel it to green production, the social democratic and green growth paths, are political decisions.

Furthermore, a level of warming is inevitable which entails, absent massive climate reparations, suffering for much of the planet, because much of the planet is already suffering. A just conclusion is to put in place massive reparations. Such spillover means the struggle to prevent climate change is over. Now the struggle is to survive and adapt to climate change, and to prevent it going from bad to very bad. The slope from bad to very bad is lubricated by the amount of oil and gas capitalism continues to burn during its transitional period. The fundamental question is, once again: whose voices count in determining the pace of transition and deciding which sources of energy use are necessary and which are unnecessary. This is a question which has no non-political answers.

THE DEBATE

Current debates about clean energy are marred by a false binary between a strawman degrowth versus a dreamland of a state-managed GND run by a progressive, ecologically minded, *Silent Spring*-toting social democratic administration that can begin to systematically plop in clean energy for dirty, moving at lightning speed to a continental grid.

The latter vision relies on physical phantasmagoria and political pipedreams. In the first place, there will be no social democratic candidate in any office in the political core outside of Iceland in 2021 able to put in place a social democratic agenda, including the demand for energy democracy. But even were that to be the case, current proposals like those of Robert Pollin, which rest on a full replacement of the current energy system with renewables, are probably not realistic and certainly not just.[19]

First, these proposals suggest current forms of energy are fully substitutable. While in principle countries could develop enough renewable energy and storage to produce the same amount of energy as they currently do, if not more, one question is how fast this can be done. If overall energy usage is staying constant or increasing, it will take longer to replace all that energy production with fresh physical plant. If overall energy usage is decreasing, it will take less time. The less time it takes to get to zero emissions, the fewer emissions are placed into the atmosphere, and the less the world suffers the effects of climate change. There can be no objective or realistic answer to these questions outside of what political forces decide is the best path. But there is abundant research showing that 1.5 or even 2°C rises are probably impossible to avoid without reducing core energy use.[20]

A second issue: at a certain point we may face diminishing social returns in certain forms of technology: what is the point of cars when there is too much gridlock for anyone to move? The notion of "energy returned on investments," or EROI, is how ecological economists and engineers address this kind of question: how much energy is used for a greater amount of energy in return. The merits of fossil fuels are very clear on that front, since digging up coal or drilling for oil takes a small amount of energy and produces a huge amount of energy. At least on that basis, fossil fuels have been incredibly beneficial, although they have also dumped a waste in the atmosphere that human societies are not organized to clean up. Renewable technologies do not merely have a smaller EROI. They also pose different questions related to pollution and trade-offs: what are the costs and benefits of massive amounts of renewable energy when huge portions of society and huge swathes of land are devoted to or despoiled by pumped water storage, redundant windmills and solar arrays to deal with peaks and troughs in sunshine and wind? We have not tried to convert the entire worldwide grid, but some studies indicate that when one includes all the costs of storage and diminishing returns on renewable energy, and if one also accounts for the fact that hydroelectric power is severely ecologically damaging in ways that wind and solar are not, easy claims about 100 percent conversion of current – and in the future, globally increasing – energy use to renewables are unrealistic.[21]

Storage is knottier, since there is no way – except mining asteroids – to produce enough batteries to store the needed power. Pumped water poses its own problems, not least the amount of redundancy needed to store energy for when wind stops blowing and sun stops shining for days at a time. Huge portions of social and material energy would have to be devoted to such a project: the engineer and energy expert Ted Trainer estimates around 3 percent of Australia's GDP, in effect eliminating "growth" and the idea of macro-economic management of economic well-being based on growth. In EROI terms, some estimate that whereas now the current energy system has a 12:1 EROI, under a renewable transition it could drop to 4:1 or even

3:1 by mid-century. Although it is not clear such numbers are directly comparable, since solar and wind often lose much less power on their way to being used, such EROIs are "well below the range of the thresholds identified in the literature as necessary to sustain high levels of development in current industrial and complex societies."[22] Since the main trump card of growth-oriented green Keynesian renewable policies is that growth allows for a bigger piece and is thus politically more realistic, such calculations suggest we are wrongly dealing with the problem, even on its own terms. A full-renewable transformation and absolute decoupling, while only using a very small portion of core GDP to finance transitions in the South and the North, all the while maintaining per capita real growth rates at 2–3 percent, does not seem realistic.

Another huge problem is the CO_2 "bump" involved in the transition: building all the new industrial-technological-energetic-renewable infrastructure will require energy and that energy cannot for the most part come from renewables. Researchers estimate it would require France's current yearly consumption to build a replacement infrastructure for the country's current energy consumption.[23] The more clean energy required, the larger the bump. Hence the goal should be reducing energy use in the core. Meanwhile, peripheral states will either receive renewable grants via climate debt repayments, or they will build out a polluting infrastructure, making the climate problem even worse (there is a third option, which is simply denying development at all to states and reducing in that way their energy use nearly to zero, a process underway in the Gaza Strip).

Furthermore, energy comes with costs. Huge solar parks, which many suggest as a chunk of the transitional energy program, require thousands of acres of open land. In India, it is notable that "open" is a political category. Such parks will perforce enclose pastoralists' common lands, which they need for fuelwood foraging, fodder, and grazing. Indeed, this appears to already be happening in the Gujarat Solar Park and the Kurnool Solar Park.[24]

The costs of such transitions are even more deranged. Such estimates are based on current prices. But the dollar-cost of raw materials that primarily come from poor Third World nations with histories of colonial depredation or recent US coups d'état, like Indonesia or Bolivia, or the neo-colonial annihilation of the Democratic Republic of Congo, reflects imperial political engineering. Cheap cobalt and other minerals from the Democratic Republic of Congo (which are part of highly disposable iPhones) has meant massive human and material degradation, people suffering with cancer, and an ongoing US-incited war, which prevents the country from exercising national-popular sovereignty over its resources and demanding a just price on world markets.[25]

The recent coup d'état in Bolivia is inseparable from Western hostility to the Bolivian government's interest in decolonization, anti-imperialism, and natural resource sovereignty in general, and its holding of lithium in particular, necessary for batteries. And copper, central to a renewable energy transition, is mined with enormous destruction: the copper pit in West Papua, Indonesia, owned by Freeport-McMoRan and Rio Tinto, produces hundreds of thousands of tons of mine waste a day, covering over 90 square miles.[26] These facts do not preclude using raw materials for industrialization. They expose Eurocentric silencing of the suffering of the South latent in notions of green Keynesian growth models. Such models and their expectations of continued GDP growth rest upon current terms of trade for primary products. Such terms of trade, in turn, rest on the reduction of the quality of lives in places like West Papua and the Congo. (On the flip side of the coin, Chile's copper mines are far less ecologically damaging, even if Chile continues to receive nugatory prices for its ore and nuggets on world metal markets.)[27] If a GND's goal is justice, it would also mean that modeling certain ways of life based on denying the good life elsewhere could not stand, and that equal exchange accords, and prior and informed consent for resource extraction, would need to be part of a global GND. Batteries for certain forms of storage would be far more expensive than they are now if those prices were not imposed down the barrel of neo-colonial price engineering but rather reflected demands for just returns to national labor, and what nations decide is a fairer price for the damage extraction inflicts on their environments. The importance of nation here is important: there seems to be no way to ensure just prices outside countries acquiring sovereignty over their resources, and then banding together to ensure just payment.[28]

There are other ways it seems that energy is not easily fungible in the ways people imagine. As the energy researcher Tim Crownshaw writes, "the manufacturing of silicon wafers in solar PV panels and advanced metal alloys in wind turbines requires a lot of high temperature heat," currently sourced from natural gas and coal. "Will it be possible to run solar PV panel and wind turbine production lines using solar- and wind-generated electricity in the future? We don't know, but there are reasons to be skeptical."[29]

POLITICAL FEASIBILITY OF OTHER PATHS

Some also suggest that those technical, social, political, and ecological objections might be right and just and worthy, but should not shape what we do, nor mark out the horizons for which we aim. Pollin, for example, argues that although global leveling of emissions and energy use would be fair and just, "there is absolutely no chance that they will be implemented. Given the climate-stabilization imperative facing the global economy, we do not have the luxury to waste time on huge global efforts fighting for unattainable goals."[30] He instead calls for technology transfers – but not climate debt payments – to help Third World nations shift to sustainable energy systems.

But there is no way to know what has "no chance" of being eventually implemented, which goals are "unattainable," and what is a "waste" of time and what is not. These are arguments that hide in the cloak of pragmatism but are really about political values. Strictly speaking, sharp and immediate reductions in consumption and immediate sustainable energy conversions are possible. Such decisions rest on the political will and social power to force such an outcome. Furthermore, there is no reason climate debt payments must be a no-go zone: they were the lingua franca of the global environmental justice movement until recently. The questions are what horizons are being aimed for; how and why one determines those horizons; whom one's friends are and whom are one's enemies. There are no climate politics outside struggle.

Indeed, Pollin and others accept the need for an acute political-social struggle: he notes that ExxonMobil and Saudi Aramco are "powerful vested interests [which] will have to be defeated."[31] The only argument against attacking even larger sectors of US/EU capitalism in order to move to a planned and egalitarian social system is that those powerful vested interests should not be confronted, a bit of a tautology. Or that confronting the massively powerful oil sector is more feasible than confronting capitalism. Such a claim makes sense if one's overall strategic perspective is based on splitting the antagonists' camp, and getting certain sectors of capital on board. Getting capital "on board," though, forgets or suppresses what capital is and where it comes from and who pays the price of getting sectors on board. Historically, social democracy rested on profits from the Third World.[32] It also relied on relocating the most polluting industries there.[33] Plans for domestic green Keynesianism are not merely bad because they do not adequately take stock of the amount of political and social struggle needed to wrest victories from the domestic ruling class. They are also blind or uncaring about where value, capital, and profit come from. In this inattention, they set the stage for ensuring that concessions to labor from US/EU capital will come from the periphery.

Against such a horizon, eco-socialism is the imposition of a different social logic onto the productive forces – the notion that energy systems should be run to balance the warring needs of avoiding ecological damage and providing necessary energy for human beings. Such goals necessarily conflict with green anti-racist redistribution, in its openly imperialist versions, as well as softer versions based on global social democracy or a global Green New Deal, arguing for managed capitalism in lieu of "hyper-globalization."[34] Social decisions about how to design energy systems so as to accommodate these distinct social needs and values will conflict. And the political forces arguing for different goals will conflict, as well.

ALTERNATIVE PATHS AND A PROSPEROUS WAY DOWN

Imagine a different scenario, based on radical reductions in US energy use, to an amount more than sufficient for a very good life, with plenty of food, excellent housing, healthcare, transport, and a modest and elegant industrialization.

How to do it? The critical sector is energy use. As technology critic, writer, and plant breeder Stan Cox points out, in the most comprehensive and globally just map for how to downshift US power needs, we need to take a series of steps. First, put an "impervious cap" on the total US supply of fossil fuel – oil, gas, and coal. Then put in place mandatory annual reductions, with the target being zero in ten to twenty years (and remember, that by itself means more warming, which means enormous damage to the planet and its poorest peoples). The government would sell permits to companies. Sell, not give. Fuel imports and exports would be banned. Furthermore, offshoring emissions would be eliminated through selling permits to importers and tamping those down to zero as well over a certain period of time.[35]

Electricity generation overall is around 43 percent of total CO_2 emissions, but as with much of the ecological impact of capitalism, responsibility sharply varies depending on wealth. In the US, for example, per capita energy use is about 12,000 kilowatts per year, and in Canada, it is around

14,000. China uses 4,470 kilowatts per person per year, and India uses 1,181. Bangladesh uses 351 and Ethiopia uses 65. The Gaza Strip uses 0.1. Africa and US warzones like Palestine and Yemen are at the bottom of per-capita rankings. South and North do not tell us everything. For example, China and India remain semi-peripheries and peripheries. India especially uses far below what it needs to give adequate human development to its people. But each countries' electricity-generation paths will have tremendous world-ecological consequences. Still, North and South tell us that access to electricity is an issue of acute uneven development on a world scale. A fifth of the world's population has no real access to electricity, and the next 20 percent uses just 2 or 3 percent of global electricity.[36] And countries like the US in effect import the energy use of countries like China, through import of manufactures which have been very largely offshored.

On the other side of the coin, buried in annexes or murky assumptions in the models for full conversion to renewable energy, is a dirty secret: the claim that some nations deserve more energy than others. For example, the highly touted models from Mark Jacobson, a professor of civil and environmental engineering at Stanford University, show how the US and the world can shift to 100 percent renewable energy. He argues, however, that South America on the whole would use 1,413 watts per person per year, 1,007 for Southeast Asia, 625 for Africa, 755 for India. The current US average is 9,500 watts.[37] In theory, such disparities will be partially softened by efficiency improvements. In practice, they are calls for permanent disparities in access to energy. Of course, those numbers assume population growth, which the North increasingly opposes – the Malthusian agenda.

DOES THIS MEAN AUSTERITY?

The North should shrink its energy use, and a People's Green New Deal would be a plan for that shift. Everyone on Earth deserves the same rights to development, including rights to food, water, appropriate housing and cultural interchange and the power levels needed to secure those rights.

Estimates of how much energy use is necessary for "the good life" vary enormously. The imperfect Human Development Index, a measure of national achievements in human development in three arenas, lifespan, education, and standard of living, is one way of assessing countries' capacity to give their people decent lives, and the resources needed to do so. The United Nations Development Program slots countries into three groupings:

those with low HDI and scores of 0–0.5, medium HDI and scores of 0.5–0.8, and high HDI: scores of 0.8–1. To achieve 0.885, Malta uses 4,400 kilowatt hours per person per year. This number is considerably higher than world per capita energy use, which was 3,081 kilowatt hours in 2018, although many future reductions in energy use should occur due to massive investments in efficiency. But Malta is a capitalist state, riddled with inefficiencies and gratuitous and irrational forms of energy use which exist by dint of it being a private property regime – having energy-gobbling systems of transport, for example.

Consider some other real-world examples of countries which use a lot less energy. Using existing technologies, Cuba and Costa Rica give something equivalent to the good life to their entire populations on the basis of nearly sustainable energy use. Cuba uses a lot of diesel because even though the island is doused in sunshine, it cannot generate enough solar energy to power countryside mass transit, in part because of the vicious economic embargo. Cuba, alone, has an HDI on the threshold of "high" (0.778) with per capita kilowatt usage far below the world average and even estimates of a "vital minimum" of 2,000 kilowatt hours. Some may say people are still poor in Cuba, and that's true: HDI doesn't tell us everything. But Cuba is also blockaded, and shows the possibility of high human development with incredibly low environmental impact.

What about hypothetical outcomes? In a modeling exercise, Joel Millward-Hopkins, Julia K. Steinberger, Narasimha D. Rao, and Yannick Oswald show that global final energy consumption could be reduced to the levels of the 1960s in 2050, with thrice the population. Their model allots people food, cooking appliances, cold storage, housing, thermal comfort, and one household computer. Often, those advocating using less energy than is currently used are implied to be evangelists for turning humanity into cave-dwellers. These, in the authors' words, would be caves with:

> highly-efficient facilities for cooking, storing food and washing clothes; low-energy lighting throughout; 50 L of clean water supplied per day per person, with 15 L heated to a comfortable bathing temperature; they maintain an air temperature of around 20°C throughout the year, irrespective of geography; have a computer with access to global ICT networks; are linked to extensive transport networks providing ~5000–15,000 km of mobility per person each year via various modes; and are also served by substantially larger caves where universal healthcare

> is available and others that provide education for everyone between 5 and 19 years old. And at the same time, it is possible that the amount of people's lives that must be spent working would be substantially reduced.

This would require a mere doubling of global renewable capacity, a very realistic goal.[38]

This might seem like so much social scientific dreaming. It is not, or it is not just that. Too often forgotten is the huge amount of waste that is built into capitalism and the private property regime. Consumer goods, for example, are not produced for people, but for profit: planned obsolescence leads to greater profit rather than planning the production of products for permanence. Remember here that hierarchical social life based on greed and accumulation prevent the human level of society from rising, and prevent humankind from using the existing technological inheritance, including the collective shared and keystone commonwealth of humanity, knowledge, for the collective and egalitarian betterment of society and its poor. We do not know, however, what kinds of technologies an eco-socialist society would produce, or what the struggle for eco-socialism might produce.

And technology is indeed very important. Energy is not just linked to politics: energy technology is itself political. Windmills are broadly capable of being built in many places, but only some places are suitable for serried ranks of industrial-scale mega-wind. However, decentralized wind has its advantages, including that it can be directly connected to industrial plants powered with compressed air, or connected to storage systems.[39] Decentralized small-scale compressed air storage may be possible to distribute on a very wide scale in lieu of the mineral extraction inherent in fabricating batteries from shiny white lithium. And compressed air systems last far, far longer than do easily exhausted lithium systems. Their major problem is density: for rural areas, this is not a problem, as there is space aplenty. For urban areas, existing technology will need to be developed quickly – a great example of how the politics of technological development is currently overdetermined by capitalism and colonialism, and how a different social system would sow an entirely different kind of people's technology.

Another possibility is retooling energy demand to energy supply. Current thinking about energy transitions rests on the notion that we should use energy in a planned eco-socialist society the same way we use it under capitalism. But there is no reason for that to be the case. Factories, which are mostly automated, can work when there is sufficient power and go idle when there is not. Goods can be transported by ships and trains, which move when there is power and stop when there is not. This is not, as I show later, an obstacle to a complex society. It is an obstacle to a complex society that looks like the one in which we currently live.

REALISM AND AUSTERITY

Finally, to circle back to realism. Talk of "realism" does not help to clarify the energy program of a global People's GND. It covers it in a pall of confusion. There are too many variables, each of which are determined by the ultimate independent variable, that quicksilver-like substance called collective human will, which never stays in any of the neat modeling containers of social democratic-technocratic pragmatism. Is it possible or impossible to create thousands of factories, operating for socially determined ends, and not for profit, and publicly owned, and in the next five to ten years? And is it possible to convince the US public that our current energy system is unjust, and a transition should aim to give everyone on the planet equal access to energy alongside equal rights to living in a clean environment?

These are all political, not technical, questions. Answers to them, which turn on what is "realistic" in a US/EU political context, reflect the values of the evaluators more than any kind of objective assessment of what is reasonable. Much of what is given the convenient label of being reasonable is about presuming the continuation of capitalism and imperialism.

In contrast, a principled program for worldwide energy democracy would rest on very high and rising redistributive carbon taxes, paid at the point of production, not consumption, choking out the coal and petroleum industries; the immediate conversion of all worldwide industrial plant that is currently producing unneeded consumer goods to clean energy technology, including energy-powered transport; technology grants (not tied aid and not loans) to the Third World to build out their renewable energy capacity; prior consultation with communities living on resource pools like cobalt and lithium; just prices for raw materials, with international legal provisions for Third World commodity cartels; the nationalization without compensation of the fossil-fuel companies; the nationalization of privatized grids; the build-out of national and international and transcontinental smart grids; local control vested in the smallest possible deliberative units over where to site renewable energy units, so that they do not end up destroying the lands of small farmers or poor people, especially in the Third

World, as is currently occurring in southern Mexico; and finally, massive public-funded research and development, funded by the First World, at least at first. As the Intercultural Dialogue to Share Knowledge, Skills and Technologies of the World People's Conference on Climate Change and the Rights of Mother Earth states, we should

> Create in each country and worldwide a bank of knowledge, with technologies aimed at reversing climate change and environmental crisis to ensure truly sustainable development that is available to all peoples of the world, being consistent that knowledge belongs to everyone not those who've been wanting to privatize it.
>
> Formation of a platform for exchange of information, knowledge and technology of free assignment, administered and maintained collectively by the people, that is, open knowledge technology in respect of the sovereignty of peoples.[40]

Such a call should be the basis for a just development and sharing of the world's collective knowledge.

4

Green Social Democracy or Eco-Socialism?

When Congresswoman Alexandria Ocasio-Cortez, alongside Senator Edward Markey, put forward draft legislation for a Green New Deal in early November 2018, it would have been hard to predict that the phrase would become the common shorthand for large swathes of ecological politics in the Global North, displacing and erasing an earlier radical environmentalism tied to demands coming from the Global South. Suddenly, long-time advocates for climate reparations like environmental journalist Naomi Klein were backing non-binding legislation that targeted a sharp reduction in US CO_2 emissions by 2030, along with boosting the US as a clean-tech leader. The legislation surfed sentiment about global warming produced by the recent IPCC statement calling for sharp and immediate reductions in global CO_2 emissions to avoid the socio-ecological disasters which a 2°C rise as opposed to 1.5°C might cause.[1]

The Markey/Ocasio-Cortez Green New Deal has become an unexamined reference point for discussion about the GND. The GND and the networks, ideas, and policies it came from, reflected, and contained, suddenly became a kind of orienting pole for progressive discussion of the GND more broadly. This happened in part because it briefly seemed reasonable – or was made to seem reasonable – that AOC and perhaps a progressive Democratic president would put in place radical climate legislation. At the same time, the Markey/AOC program gobbled up ideas and branding and political space from more radical green discourse – from the Third World, but also more radical US GNDs which pre-dated the Ocasio-Cortez/Markey legislation.

There is now a din of technocratic chatter, a blossoming of technical proposals, and a mess of blueprints for the future. A huge array of programs and political forces now vie to fill brand with substance: from just transition or eco-socialist utopia to UN scaffolding for a decaying capitalism or green mask behind which the financialization of nature can proceed with the savage impersonal coldness of carbon calculators, Clean Development Mechanism chits, and corporate quarterly earnings' statements.

Meanwhile, the role of AOC and the rising progressive, or, if one prefers, democratic socialist, bloc, and the strategy of achieving either eco-socialism or green social democracy through the equation electoral strategies + social movement pressure = radical anti-racist GND is largely unexamined. There is also little debate about how that strategy interacts with the alternative proposals pushed by journalists and intellectuals that seek, in that eternal hallucination of US politics, to "push liberals from the left," without any clarification of what it would take to push them or who exactly is to do the pushing. Clarifying the roles of different proposals, showing where they intersect, and examining the forces behind them illuminates the contours of the political battlefield. This task also helps us understand the costs and consequences of linking "systems change" to the politics of the US Congress, perhaps the most important consequence of the under-examined idea that eco-socialism can arrive through electing enough allied legislators. Before discussing political strategy itself, it is useful to examine social democracy at a more general level.

WHAT IS (GREEN) SOCIAL DEMOCRACY?

To situate the Markey/Ocasio-Cortez GND or their kin, we should note they were essentially born as green jobs-for-all programs. They crystallized in their current and highly sellable political form from a thickening social democratic ambiance across all sides of the Atlantic, as soon-to-be pummeled icons for social democracy like Jeremy Corbyn and Bernie Sanders sought left-populist redistribution, and socialism became largely identified with social democracy. While the thrashing which more conservative elements of the Democratic and Labour Parties delivered to Corbyn and Sanders are well known, social democracy in general and historically is less frequently discussed or debated. Clarity about its origins and limits, even in its heyday, is important.

We must distinguish social democracy as a bundle of policies from social democracy as historical phenomenon. As a bouquet of human rights like free healthcare, a highly functioning infrastructure, extremely low levels of inequality, and freedom of expression, social democracy appeals to many in the wealthier countries for whom life is a barrage of attacks on life, family, nature, health, and dignity.

As a historical form of capitalism, however, social democracy is a very different beast, a barely tamed capitalism which continued to hunt and feed on the periphery throughout its short lifespan, and whose offspring would return to feed on the lives of the middle class and the poor in the core as well in the 1970s. It is distinguished by four traits. One, it is a class compromise between capital and labor in the imperial core. Two, to compromise, social democracies require constant growth, in order to create a bigger pie – the larger piece for capital and the smaller for labor. Three, they survive vampirically off value extraction from the periphery. Four, European social democracies were a prophylactic against the Communist contagion then spreading amongst a devastated and war-weary European working class.[2] Each trait is critical for understanding contemporary climate talk.

First, class compromise, key to post-World War II capitalism. A compromise refers to how much of the surplus-value created by a worker they receive, and what kind of things they get for their labor beyond subsistence. This is not a neat number and is best understood either by looking at shrinking wealth differentials, or by the presence of a strong social safety net, as Europe and the US had into the early 1970s. Or comparatively, by looking at capitalist regimes without class compromises. Under colonialism, slavery, or neo-colonial dependent dictatorship, people starved to death or died well before their time. Class compromises convince the core working class to invest in the system, and to see their future as linked to its future. This is not just an ideological investment. It can occur, for example, through pensions invested in corporations.[3] People's future, their ability to be safe and relax in their elder years, becomes tied to further exploitation.

Two, social democracy was a compromise based on growth. Historical capitalism is a system of infinite accumulation of surplus-value. A firm which does not seek to grow will not attract investment and will wither and die, and capitalists expect their wealth to increase. And to try to keep workers woven into the social pact, wages should increase too. Finally, as the population increases, the size of the economic pie must grow hand-in-hand with population growth. All of these dynamics require a growing material base of the economy, or ever more things – more material used means a far heavier impact on the non-human world.

Three, because historical social democracy did not expropriate the ruling class, the value people consumed had to come from somewhere. It came largely from the periphery. The productivity of Honduran, Congolese, or Chinese workers picking and planting coffee and bananas or more recently assembling iPhones or mining cobalt goes largely to the core, where it is the object of social conflict: more profits for the owners of the companies which harvest and import the crops and phones, or higher salaries to boost core populations' consumption. But whether more of the value goes to capitalists as profit, perhaps to be reinvested in productivity-enhancing metal engines, or to workers as consumption goods, it did not and does not stay in the periphery, where wages were and remain suppressed. This phenomenon is called value transfer. Decolonization, which did much to put a stop to famine and plague, slowed but did not stop the outward flow of value.[4] The post-1970 counter-revolution against the peripheral lock-in of value explains why people there are ever further famished amidst much ballyhooed advances in agricultural productivity, what the economist Utsa Patnaik calls a republic of hunger.[5]

Four, social democracy was not merely legislated into office. And it was not merely the product of surging social struggles in Europe and the United States. It was a child conceived in fear. Marxist national liberation movements, Maoist China, and the USSR convinced the Western ruling class that some bribe to their own working classes, then engaged in radical labor struggles, was necessary for firming up capitalism as a mode of rule. Social democracy was the offspring of bottom-up revolution, not top-down beneficent reform.[6]

Two models seem possible for green social democracy. One, factories across the wealthy core retool assembly lines and machinery and churn out green tech with a massive boost from the state: made-in-America solar-and-wind, a strong possibility under President Joe Biden. Furthermore, the state could direct large streams of capital towards compensating care work and restorative work. To the extent these blueprints imagine US industry flourishing off global demand, they entail a "workshop of the world" model, a throwback to the US as an "Open Door" export platform. In such a model, they would require physical material from the remainder of the world, priced low enough so as to continue to allow core consumption and green value-added industrial production. So far, this seems to be the path, with the bulk of green jobs in the old imperialist core and China.[7] Green social democracy could also slot core laborers into services and other non-industrial labor, while keeping the US largely post-industrial. Both scenarios would play out in ecological imperialism and environmentally uneven exchange: poisoned waters in Chinese rare earth mining zones, slurry ponds pocking areas abutting mining pits. If the industry stays abroad, the core will continue to import labor-time from the periphery: capitalism making consumption and accumulation in the core dependent on eating the hours and lives of people in poorer countries.[8] Furthermore, cheap clean-tech products from export zones in southern Mexico could end up being powered by windmills that are already displacing poor Indigenous peasants, as debris from factories pollutes aquifers and cancer rates spiral.[9] Indeed, renewable does not mean socially just, with hydroelectric and other installations eliciting vast resistance from Indigenous and other peoples.[10]

These things do not have to happen. But they will happen unless they are prevented from happening through awareness of the possibility of them happening, and people preventing them through international movements which acknowledge and build on one another's demands and concerns. Social democracy's record here, particularly in the US, does not inspire trust. The key issue is historical and current value flaws, or less opaquely, the fruits of colonialism and the issue of climate debt repayments. The tactical, confused, or unintentional but consequential quietude of green social democrats on the key demand of the planetary poor, ecological debt, and not merely the much smaller issue of debt cancellation, is an omission which silences the politics of the periphery, whatever its intent. Ignoring the demands of the periphery for radical change means that capitalism and poverty will prevail there. To the extent the periphery does not restructure its own agrarian systems along socially just and environmentally regenerative lines, it will remain a literal and figurative plague reservoir, from whence inequality and pandemics like Covid-19 can infect the core. Out of humanist empathy and self-interest alike, people in the wealthier countries should wish for a revolutionary transformation in the poorer countries. For all those reasons, clarity is important.

People who are not social democrats – or even earnest social democrats who merely want a softened and constrained capitalism with good social rights for poor people – should for different reasons approach gingerly programs which swaddle a steel-hard capitalist clean-tech core in the soft papoose of eco-socialism or humanist-ecologist notions of non-commodified social reproduction, rooftop gardens, or other beguiling fragments of a green anti-capitalist program. We must inspect these formulas with care and circumspection. Both because they are unlikely to produce social democratic outcomes on their own terms, because social democracy means people continuing to suffer elsewhere, which is not a humane outcome, and because social democracy is clearly a historically fragile edifice, vulnerable to counter-revolution. Because candidates and politicians are the products of one of the industries at which late US capitalism has excelled with aplomb, advertising, we should focus on programmatic planks and not the glossy images and ghostly promises which float in front of our eyes. It is also critical to be clearer about where the idea of the Green New Deal comes from, and what that implies for global class politics in an imperialist world.

AOC AND THE GREEN NEW DEAL

Environmental politics have a history. If the global environmental justice movement, tracing from the Chipko movement in India to Indigenous ecological networks in the US to the Cochabamba statement, has long represented one strand of environmental politics, the notion of eco-modernization has represented another.[11] Thus, against deceptive claims that the notion of swirling together the "red" of labor with the "green" of environmental protection has emerged like a divine revelation from Ocasio-Cortez, the idea is old – in fact, it is a mainstay of a sector of the union bureaucracy and mildly progressive NGOs. From the early 2000s, an array of forces from Van Jones to the AFL-CIO, the labor union responsible for supporting US imperialism, have supported a labor–corporate coalition to green the US economy.[12] While proposals differ, they included calls for mass weatherization and retrofitting of homes, for "good, green jobs...hardly anti-business," and a green industrial transition.[13] There is a long history of green politics which seeks to align the interest of the unemployed or the poorly employed with decent jobs, social interests in stopping CO_2 emissions, and corporate interests in protecting a rejuvenated capitalism. Furthermore, from the far left, the US-based Green Party has long advocated a transformative GND going beyond CO_2 reduction to the just transformation of US society and its relationship to the Third World.[14] And 2008–2009 saw a bloom of GNDs meant to manage the merging capitalist and ecological crises while maintaining industrial capitalism – a bloom which withered as dollar-inflated bubbles re-boosted global capitalism, temporarily muffling demands from high and low for just transitions to a new social system (one that could be better or could be worse).

The Markey/AOC program essentially upgraded and broadened the earlier liberal programs.[15] While their blueprint has been branded as eco-socialist, alongside AOC being marketed as a democratic socialist, the substance of her politics is left-liberal. On its own terms, this is nothing new: left-liberal Democrats have been in Congress for a very long time. The danger lies in identifying an anti-racist Green Keynesian tract as a "transitional program," anti-capitalist, or related to eco-socialism. Such mislabeling can confuse anti-capitalist politics, by implying the anti-capitalist and eco-socialist movement has far more power than it does, or that the electoral path to eco-socialism is feasible. To take one example, Benjamin Selwyn announces in his strategy his *assumptions* which he does not defend, the first of which is that "a GND bloc emerges and wins political office through democratic elections in one country," based on the buzz around the Markey/AOC "Left/democratic-socialist" GND.[16]

Consider what the Markey/AOC draft legislation stated and what it left unstated. The document acknowledged historic emissions, noting that through 2014 the US had 20 percent of net global GHG emissions and ought therefore to take a leading role in the transition, adopting that get-serious soundbite, "just transition." The centerpiece is a smart grid and renewable energy, good as far as it goes, although guilty of climate reductionism, or reducing the environmental crisis merely to the climate crisis, a sin of most climate chatter. The plank about local sustainable and carbon-sowing agriculture is also good (and the fruit of an increasingly vibrant, self-aware, environmentally conscientious sustainable farming movement in the US). There is a lot about frontline communities, repairing past harm and fixing exclusions.

However, there were two paragraphs about which the silence from progressives has been deafening. First, the militaristic and Fortress Nationalist framing statements:

> Whereas, climate change constitutes a direct threat to the national security of the United States...by impacting the economic, environmental, and social stability of countries and communities around the world; and...by acting as a threat multiplier.

Talk of "national security," and "threat multiplier" is mimeographed from Pentagon boilerplate. Maintaining the Pentagon system and deploying the ideology of national security means the national insecurity of the southern nations which the Pentagon targets. It is also a gesture of conciliation to the entire national security architecture of the United States, from think-tanks to generals, from liberal nationalists to those entering and leaving the revolving door between the Pentagon and Lockheed Martin: the steel skeleton of US accumulation is going nowhere. Then there is the core of the legislation, which focused on renewables and clean tech:

> **Promoting the international exchange of technology, expertise, products, funding, and services, with the aim of making the United States the international leader on** climate action, and to help other countries achieve a [GND, which] must be developed through transparent and inclusive consultation, collaboration, and partnership with frontline and vulnerable communities, labor unions, worker cooperatives, civil society groups, academia, and **businesses**...providing and leveraging, in a way that ensures that the public receives appropriate ownership stakes and returns on investment, **adequate capital (including through** community grants, public banks, and other **public financing**), technical expertise, supporting policies, and other forms of assistance **to** communities, organizations, Federal, State, and local government agencies, and **businesses working on the Green New Deal mobilization**.

The bolds are mine, which are not meant to drown out the rest of the document but to highlight a central but under-remarked element. Its shape should be familiar to anyone who knows how slush funds, subsidies, and tax abatements work in the US: socialize costs, privatize benefits. This is a state industrial policy to shift the global energy system and industrial sector, from CO_2-emitting to CO_2-neutral tech. Furthermore, the US as "leader" is a rhetoric of nationalist industrialization, gestating in the murmurs of Biden climate plans blueprints for "buying American."

In certain ways, this was a near-doppelganger of the Australian Breakthrough's call for community–corporate–state symbiosis to carry forward a Great Transition. It was drafted, however, for a different readership in the imperial core: not the business or national security sectors where a will to power can be out in the open. Its audience was more progressive sectors, especially those charged with coopting and confusing mass movements – the historical task of the Democratic Party.

Still, the inclusion of fortress nationalist language about "national security" was not a random drafting error, the foible of a fresh congressperson. Politicians who use language of national security are politicians who do not wish to confront and usually wish to expand the Pentagon apparatus. Furthermore, the actual foreign policy positions of AOC are relevant, signposting the limits of the progressive foreign policy ambitions of the current crop of legislators. For example, in 2019, when AOC was asked if she saw President Nicolas Maduro as legitimate, she replied, "I defer to caucus leadership on how we navigate this."[17] Caucus leadership under Pelosi stated around then, "I support the decision of the National Assembly, Venezuela's sole remaining democratic institution, to recognize Juan Guaidó, President of the National Assembly, as the Interim President until full, fair and free elections can be held."[18] Can one have an internationalist Green New Deal, or an eco-socialist GND, which rejects southern self-determination?

As the writer Angela Mitropoulos points out, a politics that adopts rather than challenges the language of national security aims to enfold rather than challenge Global North nationalism.[19] This is the case even though the Markey/AOC GND considered extending economic rights to previously marginalized US populations. For it sought to do so while sidestepping how US wealth, which the program sought to redistribute, is based on pillage. Such a notion of nation was based on exclusion. It did not enfold concern for the national rights to sovereignty and liberation of peoples in the Third World. And while the program somewhat remarkably called for "protecting and enforcing the sovereignty and land rights of indigenous peoples," such calls were severed from the efflorescence of Third World solidarity accompanying the radical strands of the Indigenous movement.

Furthermore, the parts of the document demanding redistribution are ambiguous. That ambiguity is an attempt to please various social forces. Depending on emphases, the focus on public funding for private gain is not very different from the Climate Finance Leadership Initiative roadmap for mobilizing trillions towards a publicly funded and guaranteed Great Transition. Furthermore, "Frontline and vulnerable communities" are not classes. They are clay-like concepts easily molded into the framework of capitalist diversity efforts, placed in the kiln of a capitalist US Congress where they will be hardened into decoration on the imperial agenda. It's worth repeating that Senator Elizabeth Warren sponsored legislation, in some ways like and in other ways unlike the Markey/AOC document, which envisioned partnering with "Indian tribes [and]...institutions of higher education, *including historically Black colleges or universities*" for Pentagon-funded clean-tech development. Or that the CFLI suggested "Corporations and investors can also support the *just transition* of communities and workers by incorporating social criteria into their investment decisions."[20] Words and phrases like "just transition" and "community," the former forged in the crucible of the global justice movement, can become charms gleaming in the eyes of concerned citizens who mistake the word for the policy.

Furthermore, the legislation symbolically evokes yet substantively elides climate debt. For even Todd Stern, the personification of Obama's world-killing at the Copenhagen climate negotiations, accepted the US ought to symbolically recognize its historical misdeeds in the course of development: "We absolutely recognize our historic role in putting emissions in the atmosphere, up there, but the sense of guilt or culpability or reparations, I just categorically reject that," Stern stated.[21] Mere recognition of historical fact is meaningless without reparations and can stand in for them, papal indulgences paid to Western public opinion.

Two aspects of the political context into which the legislation was plopped also are relevant. First, AOC's allies and accomplices. The second, the substantive costs of such nebulous statements becoming identified with democratic eco-socialism. As the *Atlantic* noted, the legerdemain of a barwoman from the Bronx pulling herself up by her bootstraps is not how AOC came to be a household American name. Justice Democrats, a new progressive organization formed by former Bernie Sanders staffers as a US Political Action Committee, "had recruited Ocasio-Cortez and helped shepherd her campaign to victory." That group has liberal policies which fall short of social democracy: advocating a $15-dollar inflation-indexed minimum wage, for example, which would be about $31,000 dollars a year, short of a living wage.[22] And although AOC is widely described as a democratic socialist, she does not commit to that label herself. When asked if she is a democratic socialist, she said, "It's part of what I am, it's not all of what I am."[23]

Meanwhile, other evidence points to others who probably had different agendas in putting forward the legislation. AOC's former chief of staff, Saikat Chakrabarti, stated: "There are literally trillions of dollars of private capital waiting to get invested in anything that is low risk with moderate returns. This is trillions of dollars that would flow into a Green New Deal."[24] From Chakrabarti we learned, "The interesting thing about the Green New Deal is it wasn't originally a climate thing at all...We really think of it as a how-do-you-change-the-entire-economy thing."[25] In other words, the GND was born a plan for managing the economy, investing idle capital into factories for solar panels and white gleaming turbines, while keeping a big part of the US population in working poverty, and claiming greening capitalism and cleaning up the planet could go hand-in-hand. And when climate legislation is actually implemented, as it almost certainly will be in some form under the Biden administration, we will see a new "green spirit of capitalism," in the phrase of geographer Jesse Goldstein.[26]

More than anything, the Markey/AOC legislation has been important as marking the limits of legislating progressive politics, and electoral strategies to socialism with which they are bound. It is less useful then to focus on AOC, who after all co-submitted the legislation, and more useful to see her as the most prominent of a new wave of progressive Congressmen and women, carrying with them ambitions and limits. We would better understand AOC as having consolidated left-liberal agreement around a program for massive anti-racist green Keynesian industrial policy from diffuse discontent, and a rising interest in a sometimes opaque version of socialism.

It was striking, then, to have seen co-authors of a recent book on a left GND, Alyssa Battistoni, who has done valuable work on social reproduction's role in a green transition, and Daniel Aldana-Cohen, whose urban green transition thinking is also estimable, stating that Ocasio-Cortez's GND "looks more like eco-socialism than the morbid capitalism we know so well." The characterization does not tell us what eco-socialism is or what might be so reactionary as to not be eco-socialism. Perhaps most worryingly, the duo defends – rather than criticizes – AOC's elision and suppression of the pole of unity of the formerly colonized world: climate and ecological debt. Battistoni and Aldana-Cohen praise the legislation: "It acknowledges that the United States is responsible for a disproportionate amount of emissions."[27] But does that matter? Such acknowledgement was also mouthed by that quintessential emissary of climate colonialism, Todd Stern. Acknowledgment is not what people need. In accepting exclusions rather than fighting for inclusions, political horizons are foreshortened before the political battle is really joined. How can one get climate debt on the agenda without fighting for it to be on the agenda? And how can one gloss the spokesperson of a capital–labor alliance for domestic anti-racist green Keynesianism, AOC, as veering towards eco-socialism, and expect a program towards the elimination of capital to emerge from someone who is simply not calling for anti-capitalist struggle?

Labeling the AOC/Markey plan "eco-socialist" when it was not, raises questions for those who advocate an eco-socialism based on the agreements from Cochabamba: what are the costs of fusing one's ideology and program with those who advocate something considerably to the right of eco-socialism, and labeling the latter close to eco-socialism? What are the costs of allowing the word socialism to describe anti-racist Green Keynesianism? In this context, and as a testament to the tremendous confusion and disorganization dominating the discussion, it is worth noting the wider scope of compromise. For example, the Transnational Institute, a historic institution of the European progressive left, which a decade ago supported the Bolivian demand for climate debt repayments, recently called for US treasury purchase and "decommission [of] major US-based fossil fuel companies," saying, "Central banks should...purchase corporate bonds to stimulate green investment," while only in the long run "changing" – not eliminating – the "nature of multinational corporations": calls for public money to flow to private corporations so that they will cease ecocide.[28] Such programs, to say the least, do not help to develop an independent pillar of power which could impress upon the state a transformative GND reflecting the interests of the landless, ranchers, slum-dwellers, unemployed former Rust Belt workers, small farmers, underemployed, underpaid, or suffering teachers, and the Third World.

THE GREEN NEW DEAL MOVES LEFT

Two interlocked manifestos were first out of the leftist gate, structuring portions of debate around the GND: Alyssa Battistoni, Daniel Aldana-Cohen, Kate Aronoff, and Thea Riofrancos's *A Planet to Win* (*APTW*) and Naomi Klein's *Burning Up*.[29] Both were partisans of the Markey/AOC GND and the much more radical but still limited Sanders GND, which sought full renewables in electricity and transport by 2030, 20 million new jobs, massive investment in restorative agriculture and land management, and in parallel, Medicare for all.[30] Each book supported the other, through cross-cutting media appearances and endorsements. In the short run, each aimed to reduce inequality and "discipline...capital," while later on *APTW* looked to rupture with capitalism after a "last stimulus." (A third attention-getter, the Pollin-Chomsky program for a Green New Deal based on eliminating fossil capitalism, which rejected climate debt, structured some liberal-left debate but is unconcerned with socialism or internationalism. I took up some of its technical arguments in previous chapters.)[31]

Both *Burning Up* and *APTW* seemingly emerge from the radical tradition. For Klein, racial capitalism, settler-colonialism, anti-capitalism, and anti-colonialism are omnipresent. But they are an aesthetic, not a diagnostic. They linger as wisps – spectral presences hovering around poor humanity. But they do not precisely name the enemy, nor do they structure strategies to defeat it, nor do they point to comrades in such struggles. Klein nowhere lifts up radical-popular governments or peasant movements, let alone Communist parties, as potential agents of social transformation. Rather, quite unlike an earlier Klein, Klein c. 2019 saw promise for social movements and parties in the North even if they do not wish to build with the South. Part of how she massages this contradiction is to support forces that seek to reinforce green capitalism, like the AOC/Markey legislation, praising them to excess, over-gilding their radical credentials, and essentially sidestepping imperialism and climate debt.

Shapeshifting and strategic ambiguity pock the text. Markets have a "role," but are not "protagonists": people are. She ridicules unnamed critics who state, "None of this means that every climate policy must dismantle capitalism or else it should be dismissed," a position embraced by a scant few, but which allows her to sidestep criticism of the Markey/AOC GND. Yet, at the same time, she urges, we must "embrace systemic economic and social change." She calls for "democratic eco-socialism," wealth redistribution, resource sharing, and reparations. She says the top 10–20 percent of emitters need to lower emissions as fast as "technology allows." However, the power to consume and more importantly private ownership over the means of production, not whether or not our current technology is adequate, is the main obstacle to reducing CO_2 emissions from the world's wealthiest. More specifically, she suggests confronting the fossil fuel sector, keeping the carbon in the ground, keeping an eye on the central role of the US military in driving up emissions, and describing debts owed by the rich to the poor.

Much of this program, at least, is alluring, although salted with excessive baiting of those she sees as to her left. The upshot comes when Klein shifts registers – from the pulpit and its seductive generalities to the technocrat who can get pinned down on little details with their big impacts. Klein proposes a quite small 25 percent cut in US military spending. She suggests a 1 percent billionaire tax – far below average internal rate of returns on Fortune 500 companies. And while she endorses democratic socialism, she softens the diamond edge of class struggle by calling for tending to the "top priorities of the most vulnerable workers and the most excluded communities." Concern for the vulnerable, caretaking through just-transition, and talk about communities is putty, attaching as well to humanist-enfolded liberal capitalism or to green Keynesianism, as to calls for a Great Transition from Morgan Stanley and the Climate Finance Leadership Initiative.

Whatever the intent of such words, they find their place in a world whose cognitive terrain is structured by capitalist ideology and class power. The "top priorities" of the poor and the colonized are often – at least when organized – to decolonize, to secure mammoth climate reparations, to overthrow capitalism. But cottony phrases like "most vulnerable" and "most excluded," tell us nothing about class in the North, let alone North–South differentials. Are the "most vulnerable workers" the anti-capitalist South African metal workers or former Boeing mechanics or a future syndicate that brings the two together? And would not the first set of workers also want climate debt repayments, not merely a just transition in general terms? Are the most excluded communities Adivasi forest dwellers or occupied Kashmiris or for that matter Yemeni children who need an awful lot more than a mere 25 percent cut in the US military budget so that fire ceases to burn down their world? The seeming inclusivity of Klein's book sidesteps that journalists are in no position to lasso together forces with antagonistic agendas, or if they try to do so, they cannot produce analysis which is helpful for political struggle. As Angela Mitropoulos points out, "It becomes possible to bundle together these presumably contradictory ideas together because, once shorn of the material practices that would render them irreconcilable, they simply become another value-point in a spectrum of investments and, thereby, a means of portfolio diversification and hedging."[32] The Klein roadmap lacks a social agent, a subject. Who is to carry out this massive program of change, and demand to be included in such a blurry program?

Opacity is a feature, not a bug. To a large extent, people may see in this journalism what they wish. When it comes to specifics, Klein praises the "social movements" in the core declaring a people's emergency: The Sunrise Movement, Extinction Rebellion, the People's Climate March, and 350.org. However, it is a stretch to simply identify these forces as social movements. At best they are largely hybrid NGO movements wherein the top leadership loosely sets priorities and local chapters implement them as they wish. Nor are these groups accountable to working-class constituencies. 350.org has received over one million dollars from the Rockefeller Brothers Fund over the years. It also takes money from the European Climate Fund, whose mission states, "As a free-market democracy and the world's largest single market, Europe is a key laboratory for innovative business and progressive social reform."[33] Other funds come from the Arkay Foundation, which funds liberal/capitalist NGOs like the Sierra Club and Greenpeace.[34] Nor is 350.org anti-capitalist. Bill McKibben of 350.org recently stated, "But there now seems the real possibility of concerted action across the federal government to make sweeping change" because of the Biden administration's climate team.[35] Biden's actual climate plan aims for net-zero by 2050 – far, far too late – and urges we "Invest in the climate resilience of our military bases and critical security infrastructure across the U.S. and around the world."[36] As for Extinction Rebellion, it stated in September 2020, "We are not a socialist movement. We do not trust any single ideology, we trust the people, chosen by sortition (like jury service) to find the best future for us all through a #CitizensAssembly. A banner saying 'socialism or extinction' does not represent us."[37] Such organizations are not anti-systemic. They are more of a cautionary tale of NGO containment, capture, and channeling of massive popular unease. It is incoherent to at the same time identify oneself as fighting for "a new form of democratic eco-socialism," and lean so heavily on explicitly anti-socialist or pro-Democrat individuals and organizations.[38]

And while Klein is aware the New Deal was born of worldwide Communist insurgency, she concedes the current moment lacks that impulse, although there are "signs of strength": the Sunrise Movement, the School Strikes for Climate, movements against mass incarceration and deportation, or fossil fuel divestment. These struggles are important. But to compare the anti-racist campaigns to which she refers, or student-led divestment referenda, never mind ideologically diffuse, NGO-led, nascent, or disorganized formations which largely lack international impulses (the movement against mass incarceration partially excepted) to the Communist Party-linked and led hydra of the 1930s is of little value. It is, in fact, misleading. There will be no social democratic GND unless radical forces are mobilized towards much more liberatory horizons, and we know this because no political force is able to achieve its program even when it achieves state power, and there is zero prospect in the short run of social democratic green forces taking state power (radical processes in the Third World like the Bolivarian movements in Venezuela are, in fact, dismissed as "petro-populism"). Similarly, Selwyn dismisses all radical Third World forces as "authoritarian," while praising progressive foundation-funded NGOs.[39] In dismissing concrete Third World forces actually fighting against capitalism, wealth concentration, and imperialism, are these positions so distant from Ocasio-Cortez's lack of interest in supporting the *proceso* in Bolivarian Venezuela?[40]

Furthermore, many such NGOs interact with the capitalist Great Transition like dimples on a golf ball. Those depressions ensure the Great

Transition can cut through a thickening atmosphere of left sentiment, and prevent it from stopping a hierarchical transition and imposing a pro-people program. If Klein, or Selwyn, name such organizations as their social movement, how can they claim to be fighting colonialism or capitalism? This is not about making the perfect the enemy of the good but the much simpler question of agreeing on what to fight for and how to fight for it, and identifying who is fighting for what by their own self-descriptions.

If Klein is content with gestures about the GND program, *A Planet to Win* thoroughly and eloquently describes sectors that should be grown and better compensated, from teaching to healthcare to public spaces and public luxuries to affordable and alluring public housing. Understandably, the document has informed the Democratic Socialists of America's climate strategy. The authors propose trusts to remove land from capitalist valuation, cooperatives to increase the power of labor, and utilities to become clean and worker-owned. In the short run, there would be massive redistribution and higher wealth, inheritance, and upper-level income taxes, alongside "worker co-ownership of large private companies." After the "last stimulus" there would be a break with capital.[41] In fact, the book looks closer than either Klein's or the Markey/AOC document to a transitional program: an ambitious proposal going beyond what any domestic capitalist could possibly tolerate and resting on massive taxes on the rich to enrich the lives of the masses. It is by far too radical for a capitalist Congress to implement.

Indeed, much of the manifesto is more or less unimpeachable, and looks a bit like the Sanders GND, although in a few ways less ambitious. As short- to medium-run strategies, climate campaigners of the radical left can agree with it. But I focus not on what accords with eco-socialism, but what does not, because clarity comes through disagreement and conversation.

And as part of an actual program of rupturing with capitalism, this program runs into obstacles. The primary obstacle is that strategically, it is basically a program for an electoral route to eco-socialism, with social movements applying pressure from outside the state. History has given us a test of electoral routes to social democracy under present conditions. The non-socialist but social democratic Sanders and any other vaguely anti-systemic politician, as with SYRIZA and Corbyn, has so far faced bridling or breaking (breaking, so far, seems the preference). The relationship between short- to medium-term green Keynesianism and long-term eco-socialism is politically incoherent, since European social democracies' and post-war US governments' desires to steer a middle path was only motivated by their awareness of the narrowness of the political strait they were navigating, between the Scylla of unbridled capitalism and the Charybdis of Communism. Capitalism if unsoftened would drive social movements to grab the tiller and move to communism. The political vessel in which accumulation was secure would founder. This does not mean it is bad to have anti-racist green left-liberals in office. It means that they will not implement eco-socialism unless massive movements and parties outside the state, and worldwide, are fighting for actual eco-socialism, which Ocasio-Cortez, 350.org, and for that matter the Sunrise Movement, are not.

The authors' orientation to growth shares these problems of speaking to multiple constituencies, not in the interests of working through hard problems to arrive at a principled synthesis but sidestepping fundamental questions of just transition: distribution and production. Aronoff et al. deploy unfortunate strawmen against eco-socialists from engaging with degrowth: "Who will march for green austerity?" they ask. But which serious parts of the environmental movement march for green austerity? Degrowth, which is the closest to doing anything like that, does not really call for austerity at all. Rather, it has powerfully put the question of material impacts of capitalist production on the table, even if with a sometimes iffy analysis of capitalism. The question anyway should never be framed in terms of austerity, a term out of the capitalist lexicon. The question is which sectors should grow, how to decommodify them, and which sectors should shrink and how they should shrink, with an awareness that northern material use must shrink considerably through eliminating overuse and waste of global resources.

That does not mean the lives of the poor should suffer. Rejecting capitalist organization of healthcare and medical training would lead to a better, more knowledge-intensive health system. Free universal healthcare, with a far stronger emphasis on nutrition, prevention, and primary care – which most degrowthers advocate – is not "austerity." Rather, it could exemplify how decommodification and socialization could go alongside a sharp reduction in the physical resource use and GDP measure, and still improve qualitative outcomes (an issue I will develop in considering what it means to call entire sectors less material-intensive without considering the international political ecology of those sectors). As I show in the next chapter, macro-scale healthcare outcomes have not that much to do with costs, or the goods used in healthcare. Similarly, the number of calories available to poor people in the US and the nutritive breakdown of the US food supply is a question of distribution and the quality of what is grown, not of whether there is "enough." Words like austerity do not help us understand the social problem of hungry people and the ecological problem of unsustainable farming practices in the South and the North, nor the imperialist-imposed distribution of hunger. And as Aronoff et al. gracefully show, austerity is not a lens which brings into focus the distinction between luxury communal public transport and decentralized profusions of private cars which cause boundless and burgeoning traffic jams, emissions, accidents, cancers, early deaths, and foul blooms of ecological devastation across ore and manufacturing zones across the Third World.

Coming to technology, it is then not so much wrong as not helpful to state, "We see no reason to arbitrarily decide in advance what technologies will ultimately be sustainable." First, the authors do decide in advance that mass individualized electric transport is unsustainable: it does not make sense for an eco-socialist society. Equally, other people elsewhere argue that other technologies likewise do not make sense for their own march to eco-socialism. Bioenergy with carbon capture and storage ought to be campaigned against because the movements of the poor in the Third World like La Via Campesina demand it. They are right to resist technologies that assault their lives, and which are not needed for prosperity elsewhere.[42] Are they not our comrades?

Much of the problems of Aronoff et al.'s approach to technology lies in that, as they admit, the book does not deal with agriculture and land use, where the costs of technological gadgetry will fall on the South, and where the South has vibrant movements whose demands must mold any northern GND. Furthermore, agriculture is a field where dunderheaded technofixes that would mean apocalypse for the Third World seem to bloom like mushrooms. We are only discussing lab-meat within leftist spaces because it has been oddly evangelized by northern vegan-promoting progressives-of-sorts like Jan Dutkiewicz, Astra Taylor, and Troy Vettese, who call quite explicitly for a "post-meat age," complaining that livestock "gobbles up 40% of the world's habitable surface" – primarily rangeland, which would in fact otherwise be occupied by other herbivores (their scope is the world, not just the US). Of course, calling for a post-meat age means actually changing the lives of the hundreds of millions of pastoralists and integrated pasturing-forestry and smallholder-livestock systems across the Third World and replacing them with "large-scale, public-directed investment in both plant-based meat alternatives and cellular agriculture (ie, growing animal tissue from stem cells), which would expand scientific research and employment while spurring a transition to animal-free protein."[43] *APTW*'s call for R&D into meat replacement rather than rotational-intensive grazing echoes rather than distancing itself from such catastrophic injunctions.

Most importantly, we have to keep in mind that the problem is not meat, but certain kinds of meat production: capitalist concentrated animal feeding operations. These are distinct from grazing, which mostly occurs on non-arable lands and can restore carbon to the soil while improving its health and providing reasonable quantities of meat to core populations. While industrial meat production may propel Americans into endorsing veganism, people in the imperial core especially should be aware that elsewhere meat production does not look like it often looks in the US, nor is meat production everywhere so destructive and industrialized as it is in feedlots with their manure pools and maltreatment of animals, nor is US meat production homogenous.

There are also even more explicit or excluded internationalisms. Whereas capitalist Great Transitions are greenwashed modernization theory, social democratic green internationalism faces squaring inclusion and exclusion. The authors argue a radical US Green New Deal can connect with movements in the periphery through "solidarity with the communities" mining for renewable inputs. The authors call for "new solidarities and partnerships with social movements and governments around the world" – but on what basis and with which governments and to which ends?[44]

While the overall sentiment is humanist, and we should indeed fight for supply-chain justice through labor internationalism, other elements of an international just transition are more important. What is missing in the First World left is not an abstract commitment to solidarity and partnership but a committed internationalism which takes the anti-systemic struggles of the periphery as the fundamental departure point *for* solidarity. The phrases "climate debt" or "ecological debt" do not appear in the book's index. One sentence stating, "We would also prioritize mechanisms to transfer funds and technologies to the countries of the Global South to help them cut carbon emissions and adapt to climate change," does not really do justice to the anti-colonial and anti-capitalist clarion of climate debt. Near omission of climate debt is particularly harmful since each and every serious northern proposal effectively imagines the fast phase-out of all fossil fuel use. Since 80 percent of fossil fuels are located in the periphery, in the absence of climate debt settlement, such a phase-out in oil-dependent countries like Venezuela, Bolivia, and Iran would lead, in the words of economist Keston Perry, to "massive economic and social displacement and chaos."[45]

Similarly absent is recognition of the need for national sovereignty as a *sine qua non* of peripheral development, which imposes a corollary on northern struggle: not the sentiment of solidarity, but anti-imperialism as solidarity in action.

CONCLUSION

There are four problems with green social democracy. First, it is not achievable through current strategies. Two, even if it were possible, it would be imperialist and rest on devastating the South. Third, it is being marketed as something it is not: namely, eco-socialism, or the conversion of the core and the world to non-commodified and non-hierarchical self-managed social and economic relations, with convergence between the core and the periphery, and permanently sustainable scientific management of the environment. Four, it limits our political imaginations.

First, to point out green social democracy is not possible through current ways of struggling for it is simply to clarify how to force ruling classes to accept decreases in their power, or how social democracy was historically achieved. Those rights were won by a global struggle. US Fordism and Swedish social democracy were implemented amidst the immense prestige, popularity, and power of Communist states, national liberation movements, and domestic Communist forces, including the Communist partisans who were the spear's tip of anti-fascist resistance in Italy and Yugoslavia. Such forces were the red specter which pushed the core to extend social democratic economic rights at home.

Such struggle is now scant. The anti-systemic states of Latin America are under siege. Communist insurgencies in South and Southeast Asia face US-backed death squads. In the core, fascists are more organized than anti-systemic radical forces of any stripe. A few social democratic legislators do not change this tableau. There is no domestic or foreign social force capable of fighting for green social democracy, despite the understandable popularity of domestic green social democratic programs. Even the Sanders campaign was far from achieving electoral success. When it appeared to pose any kind of threat, the capitalist class consolidated and eliminated him from history's stage. Furthermore, even were Sanders to have been elected, his agenda would only have been possible to implement amidst tremendous pressure in the streets. So why not organize and write and think to build that pressure?

Two, the green social democratic program is imperialist. This is simply a question of material and where it comes from: the lithium for Sanders's proposed program for millions of electric cars, for example, could only have come from enormous resource extraction and unequal exchange for the lithium of South America. The tremendous under-emphasis on climate debt in *APTW* or *On Fire* is symptomatic of social democracy's unwillingness to engage meaningfully with colonialism and imperialism. Some authors' attacks on Venezuela and Zimbabwe, which saw major anti-racist redistributions of wealth in the dawn of the twenty-first century and blocked the ecological imperialist Copenhagen accords, are not an internationalist politics. Furthermore, an open mind towards biofuels is curiosity about an overtly imperialist technology. Perhaps worse is the curiosity towards lab-meat and compulsory global veganism, which could only mean by hook or by crook preventing people in the Third World from herding and raising animals for meat. What is to be done with the hundreds of millions of people whose lives Aaron Bastani and Jan Dutkiewicz do not find ethically appropriate?

Three, supporting Markey/AOC's GND, or more dangerously labeling it as eco-socialist, misdirects and disorganizes thinking and potential practice around eco-socialist transition. Words which stand in for the world we want to see cannot lose their meaning. We want eco-socialism: non-commodified access to fundamental social rights alongside permanently sustainable production and consumption practices and the elimination of the profit motive and any serious social hierarchies. The Markey/AOC program is not any of those things.

Four, to get anything anywhere close to eco-socialism, one must have a movement which is fighting for eco-socialism (or degrowth communism, ecological Leninism, *buen vivir*, and so forth). If one is building a struggle that accepts a compromised horizon, one is not going to reach a further horizon. Nor have struggles for communism reached all of their objectives. For that reason, it is important to be clear that much of what is proposed as eco-socialism is green anti-racist social democracy, blighted by its failure to mark out an independent and autonomous social and political path reflecting the interests of poor humanity. So what would an eco-socialist People's GND look like?

PART II
A PEOPLE'S GREEN NEW DEAL

This part includes three chapters laying out a loosely connected set of points and contributions to the conversation about a vision of a People's Green New Deal, and a conclusion with some comments on political strategy. It is good to have a reminder here of the ambitions and modesties of what I mean by vision. I don't mean an exercise in drafting the most detailed blueprints for another way of life. These chapters are instead an effort to stretch the imagination. The final chapter in particular is less about vision and more points to red lines drawn by the periphery and those internally colonized, lines drawn by living and breathing – if struggling for air – political and social struggles of the dispossessed and resisting peoples of the planet.

The chapter on industrialization, manufacturing, transport, and planning is the most explicit about vision. My obvious debt is to Colin Duncan. But Duncan, knowingly and unknowingly, was building or repeating the discoveries of a strain of British socialism linked to William Morris on the one hand, and on the other, the line of thought extending from revolutionary China onto the Arab (especially Tunisian) and Latin American debates about alternative technologies and appropriate technologies. While capitalism is tightly tied to the climate crisis and the broader ecological crisis, industrialization with its ineffable tendency to produce waste and to disrupt the environment needs to be tamped down and boxed in. We need a controlled industrialization. To call for controlling industry is totally different from a world without industry. I emphasize there are many times and places where non-industrial goods, or a heavier use of natural fibers and materials, would make us happier and more content, pollute the environment far less or not at all, and have an affinity with building a more decentralized eco-socialist society. At the same time, a familiar list of heavy industrial products, above all in the transport sector, are required for any just transition.

The chapter on agriculture is the most blueprint-like. Why? First, industrial agriculture is a horror on its own terms. Yet this is perhaps under-known even by many eco-socialists, let alone more broadly. Furthermore, humans do not need industrial agriculture to produce enough food and other materials for us to live good lives. Second, industrial agriculture is closely connected to a bevy of environmental crises, from the climate crisis to the biodiversity crisis to the crisis linked to soil erosion and damage, to pesticides and fertilizers and their byproducts leaching into groundwater and into seas and oceans. Third, agriculture is ridiculously easy to fix. Maybe the easiest to fix at technical and planning levels, through large-scale agrarian reform in most every country on the planet, nationalization and dismantling of agriculture companies, guaranteed purchase prices, production quotas organized regionally or nationally, emphasis on food sovereignty within each country or region, guaranteed living wages for rural producers and those elsewhere in the food system, and state support for a transition to agroecology. The obstacles are political. We do not need to invent new things or even test for proof of concept. We just need to do what needs to be done and what we know how to do. Fourth, agriculture is the periphery's path to convergence with the core, by producing more of what it needs locally, protecting its own environment, and ensuring large-scale domestic prosperity by distributing the land and increasing rural buying power.

The chapter on the national question and anti-imperialism shows how a Green New Deal needs to be built in a specific way, fully attentive to and heeding the demands of the South, in order to build a world big enough for everyone. This need is structured around two basic demands: one, the payment of climate debt. Two, respect for state sovereignty and, in situations of settler-colonialism, national liberation. This is not a question of tacking things onto the eco-socialist agenda, but the necessary foundations needed for building eco-socialism: how are Arab countries to even contemplate building eco-socialism when occupied by Israeli or US forces? I highlight specifically political levels of action, and how all sorts of peoples need their political right to determine their own path respected in order to carry out the task of constructing eco-socialist civilization in their parts of the world.

5

The World We Wish to See

The problem of transition to a socialist society has bedeviled generations of planners, politicians, and revolutionaries. The ecological crisis has layered ever more complexity onto the problem. The forces and relations of production have produced a devil's brew – industrial colonial-capitalism – which threatens to collapse the ecologically fragile space within which humans have built up interdependent yet savagely hierarchical and murderous civilizations. Socialist and non-socialist utopians, South and North alike, have thought plenty about the world they wished to see: from Ebenezer Howard's Garden Cities to Ursula Le Guin's endless utopias, anarchist and otherwise, to Lewis Mumford's democratic technics to Ivan Illich's convivial technologies, and from *hind swaraj* manufacturing renaissance to Ismail-Sabri Abdallah's basic needs approach in the Arab world.[1] Each blueprint has something ineffably local about it, while nowhere rejecting global interdependence. Each sketch has something artisanal while not rejecting industry. Each future finds something to cherish in the old while not lapsing into antiquarianism. And each vision extols neither bucolic idyll nor metropolitan techno-futurism even while seeking to smudge and smooth away the hard lines of a town–country divide.

At their best, thinkers like Samir Amin, Abdallah, and the Gandhian economists theorized underdevelopment as based on drain and uneven exchange. They charted a new path for development on the basis of such economic cartography: auto-centred development and delinking, or subjecting national economies' relations with the rest of the world to a social logic which reflected the interests of the poor, and not "comparative advantage."[2] Most planners sketched utopias based on a still rural world. Few have been meant for the First World, which while carrying out endless accumulation has benefited from unequal grabbing of planetary production, reckless scattering and dumping of capitalist industrialization's debris, and gratuitous and ecologically blind consumption. A lopsided world makes for lopsided politics: recurrent imperialist deviations in social democracy.

It is tough to knit together political formations from the South with those in the North that can see the long-term interest in delinking from cheap access to tropical foodstuffs, minerals, labor, and atmospheric space. Eurocentrism can prevent people from grasping that unless the Third World has the right to develop, the ecological problem cannot be solved on a just basis in the First World.

Social democratic electoralism, now partially dispelled by the fast work the Democratic Party machine made of Bernie Sanders, has been one way to imagine such change without the slow build-up of political and social power and organization that would otherwise be necessary. It should be clear – at least until 2022, when we can anticipate a bout of profitable amnesia and ludicrous enthusiasm for the latest messiah descending from Mount Grift to occupy the left wing of the Democratic Party – that to grab the tiller of national northern states in the short term is a pipedream. If Green Keynesian New Deals have been part of that intoxication, what does a more sober yet more utopian account of change look like? What are some elements needed for a People's Green New Deal? One which points towards a liberatory horizon but does not rely on the *deus ex electione* of voting in social democracy, which can provide the basis for unity in difference and for broad fronts that can evade the tar of mainstream politics?

What follows are some notes to that end. This is not a political program or a blueprint to change the world or a summary of the social forces required to do so. Rather, it is a set of analytical notes which may be helpful for people as they decide how to draft or modify programs and how to combine political forces.

In the short to medium term, and in the North, action needs to be autonomous from national politics and the two-party system. And there needs to be convergence-in-diversity around elements of a shared program. In the slogan of the Black experiment in Mississippi, Cooperation Jackson, we need to "build and fight, fight and build."[3] Such a front needs to work locally for what its people need and how they can produce and resist, nationally for hard and lowering caps on hydrocarbon use, and nationally/internationally for ecological debt repayments and respect for sovereignty of the Third World, prising open the political space for development. Ecological debt repayments and technology and knowledge transfer cannot be separated from that struggle, for they ensure that the right to the good life is shared on a planetary level.

As the to-and-fro of municipal, national, and international struggle escalates, quotas of CO_2 and burnable hydrocarbons will shrink. Technological, engineering, planning, and agricultural paths based on dumping waste out into the shared sky will be remade through collective democratic decisions, and it seems likely such processes can only come locally in the metropolitan core. What do fighting and building look like when the core's working class is caught in the miasma of a toxic and capitalist-built infrastructure and ecological cul-de-sac, where housing is irrational, transport individualized and bewilderingly inefficient, goods production simultaneously highly productive yet making oodles of crappy kitsch, which people buy because it is imposed upon them, and people overworked yet underemployed and underpaid?

A front can only build from existing strengths: the already-existing ecological society in the interstices and shadow-zones of colonial-capitalism, arenas which can or do rely less on imperialism for their social reproduction. Such a front must be wide enough to enfold petty bourgeois farmers and Black nationalist agricultural endeavors fighting for food sovereignty; red ranchers sowing CO_2 in soil; the Chicano rural proletariat; Indigenous groupings fighting for Land Back and against mineral extraction scoring and blighting their land; re-localizing manufacturing through fabrication labs and factory takeovers; endogenous development brigades in Appalachia; and the tremendous unpaid work of social reproduction in households (overwhelmingly done by women).

At every step, transitions must empower local labor, and remake development on popular and ecological lines, what the Palestinian development economist Adel Samara calls development by popular protection.[4] They must build autonomy and decentralized power into the transition. Technology matters. In the words of the philosopher Ivan Illich, we need specific kinds of tools, not merely any tools, but ones that "foster conviviality to the extent to which they can be easily used, by anybody, as often or as seldom as desired, for the accomplishment of a purpose chosen by the user."[5] Each remade trolley and bicycle network, local farm, repair workshop or carpentry site, artisanal cathedral, bicycle, localized city mega-block, internal linkage, even each power plant built, rejiggered, or reworked means increased popular/working-class power, and decreasing monopoly rule. With suitable tinkering and freshening up, blueprints for auto-centered development still make sense for the Third World, and even the First World, whose own "inner" peripheries have experimented with endogenous or inward-looking development: building local capacities and local economies, safeguarding land from poisoning and people from exploitation.[6]

NEW LABOR FOR A NEW ECONOMY?

Labor is central. Stefania Barca argues that ecological transitions should aim for democratic and worker-controlled production wherein workers also control the surplus and decide collectively where it goes, "from the workplace to society at large."[7] Control of surplus is vital so that it does not flow towards reinvestment in polluting industries. Worker control and ownership over enterprises through cooperatives is one way to do so, socializing surplus towards joint deliberation over investment, with an eye fixed on the ecology. Industrial plants can be taken over and redirected to renewables or other grassroots Great Transition infrastructure. Consider the Vio. Me self-managed factory in Thessaloniki, Greece, where workers took over the shop from the bosses and started producing ecological cleaners, a kind of lighthouse, operating within the market but offering a way of moving beyond it.[8] Of course, better in the long run is public control over what is produced, because some people are excluded from the workplace.

Another approach to placing wealth under democratic control is indirect: wage increases alongside reduced working weeks. People expand their own realm of freedom, walled off from capitalist exploitation, while throwing a wrench in the treadmill of production where increased productivity leads to more production and an economy of accumulation rather than sufficiency and time affluence, since decreased work hours means decreased overall production and therefore decreased environmental impact.[9] To ensure workers gain the benefits of increased productivity, historical and present, people can fight in the here-and-now on dual fronts: mandatory workweek caps of 20 hours or less, and massive increases, quintupling or more, of minimum wages. Social wealth will work its way into workers' hands and homes.[10] Of course, a revitalized union movement is absolutely essential to such a project, as is attention to climate debt to ensure that as more wealth concentrates in the hands of workers, it comes from bosses and owners, and not from the people of the periphery.

And there are other important shifts. Visions of a post-industrial economy are mirages. They appear in the air for a specific reason. Industrialization occurs out of sight of the North and often out of mind, in polluted southern cities and semi-urban manufacturing zones. Yet the rise of service and care work is real, a partial shift to a more knowledge and attention-based economy. Ever larger portions of work in the wealthy West are found in sectors like the university, secondary, and primary-school teaching, childcare, and healthcare – social reproduction.

But social reproduction theories seldom study the forms of very concrete production that lie beneath it in the core. We should pay close attention to what it means in terms of physical and ecological impact when we say there is a shift to services. In the United States, in part because of the immense carbon footprints of hospitals and pharmaceuticals, healthcare accounted for around 8 percent of US emissions in 2007.[11] Such rampant pollution produces notoriously poor healthcare outcomes per dollar spent, the fruit of hardwired capitalist inefficiency. Cuba's prevention-based and primary care centered holistic healthcare achieves outcomes the same as if not better than the US's.[12] Since Cuban emissions per person were in 2007 around 1/12th those of the US, it is clear that healthcare in the US or anywhere can be decarbonized with ease. And outcomes would be superb. They would be even better as restoration corps scrub the soil, sea, and air clean of carcinogenic toxins, as we learn to remediate such poisons and pull them from the landscape. And although we should embrace scientific advance in healthcare, from regrowing knee cartilage to organ transplants, the obstacle to world-class worldwide universal healthcare is not technological. It's social. It demands a social revolution to shatter the capitalist organization of healthcare and to restructure it as primary care-centered, preventative, and decentralized.

Such a shift has South–North dimensions, too. Consider staffing. As the healthcare expert Salimah Valiani points out, US hospitals increasingly imported nurses in the 1990s and 2000s, as part of a class offensive against the gains of domestic nurses, and amidst rising costs due to mounting deployment of technology in medicine. The US came to be a major employer of Filipina nurses.[13] Such nurses are reared on the land and labor and wealth of the Philippines. Their childhoods, feeding, time spent learning to read, housing stock, and medical training are part of the Philippines' CO_2 emissions, not those of the US. Such human flows occur through political and economic channels hewn by colonial violence: The Philippines is historically and now a US neo-colony with wages depressed in part because of the massive super-reserves of labor in the cities and countryside, the fruit of a scorched earth counterinsurgency against agrarian change.[14] Social reproduction in the North is provided by extremely concrete southern life and land, and woven into an imperialist world-system. Importing nurses means US healthcare development relies on the underdevelopment of Filipino healthcare: the US has 15 nurses per 1,000 people while the Philippines only has five.[15] Greater US resources would have to be devoted to raising the human beings who become nurses if they were no longer immigrants within a world-system of unequal labor exchange.

We also need to reclassify what kinds of labor count. For example, the 1970s Wages for Housework movement demanded that women receive payment for the home-bound and made-invisible work of social reproduction: cooking, cleaning, raising the future generation, and caring for the elderly.[16] Such work is essential in the production and reproduction of the use-value of human labor. Strikingly, such labor, which can loosely be placed under the umbrella of care work, is often attention-intensive and thought-intensive but not necessarily ecologically intensive. Furthermore, it is notable that almost no one suggests that tending to children and the elderly ought to be made the province of fully automated luxury nannies and nursing robots.

Part of a People's GND, in the long run, is ensuring that such labor is registered and compensated in municipal labor registries where people receive direct welfare payment for ordinarily non-compensated contribution to human society, in effect socializing such work by ensuring that the budget, the only instrument for collecting human allocation of resources in the Western capitalist oligarchies, ensures that care work of every kind receives its due. Or through direct national payments, and not at the casually degrading minimum wage, but at five times that. But as we will also see, paying labor its due must take place on an international scale, through building with, rather than at a distance from, or over the voices of, resistance movements in places like the Philippines and Cuba. We cannot have a world where care work is compensated justly in the domestic sphere while continuing to suction care workers from desiccated neo-colonies.

A NEW URBAN PLANNING

Because cities increasingly sweep across social and economic landscapes, a People's GND needs a way to deal with urban forms sprawling from cities to suburbs to exurbs, melding into undulating urban waves: New York-Newark, Los Angeles-Long Beach-Santa Ana, Beijing-Tianjin, Tokyo-Saitama-Yokohama. Such huge urban formations are testament to unequal exchange on a mass scale, metabolic rifts ripping through the socio-ecological fabric, and excessive industrialization. Such over-industrialization is uneven. Some places on the planet lack for electricity. Elsewhere, dozens of daily forty-minute flights arc from New York City to Boston and Washington, DC. Such transport reflects parochial sorts of planning rationalities based on a painstakingly engineered world price system.[17] This contrivance allows such technologies to festoon an increasingly networked set of international cities and their upper-class residents while others are deprived of social wealth, or even of states.[18]

Eco-socialist urban planning does not mean shipping people off to the countryside if they are not willing or do not know how to make their lives there. (Of course, there are those who are willing: consider Ricardo Jacobs's work on city-dwelling pastoralists in South Africa, who want land for agricultural avocations. Similar dynamics span African cities. In 2005/2006, a third of Black South Africans had a desire for land for agriculture, and a third of those were city-dwellers.)[19] So rebalancing city and countryside in the periphery through re-peasantization and re-localization is possible: above all, it depends on how relatively attractive life and work is in the countryside versus cities, even while blurring the differences between the two. Indeed, life needs to be made excellent in both, but part of that means compensatory investments in smaller urban centers and towns. To a smaller extent, this shift is possible in the core: the ongoing enchantment with community gardens suggests people would like to be more involved in doing certain kinds of manual labor, hands touching dirt and grass and trees, than they currently are. And cities pulse, growing and shrinking based on overall patterns of state planning: US suburbs and exurbs are relatively new.

To say first, cities should be re-engineered in ways that do not erase that which makes them cities: their density. Planning could take an eraser and soft grey pencil to hard black and white lines which mark off the human-engineered concrete-stone-metal land of the city from the bucolic or even human-free natural realm of the wild or the countryside. Renewed urban planning means increased public access to living spaces within cities: parks as public affluence, planted with macadamia and pear trees everywhere possible, to supply shade, to regulate temperature in summer and winter, to absorb flood water, and to convert CO_2 to wood, bark, leaves, and roots. Reworking drainage systems to channel water towards plantings and through permeable pavement will be enormously labor-intensive and would more smoothly enfold cities into ecological cycles. In more arid zones, cities ought to learn from places like Tunisia's Jerba, where cisterns lie under houses and the architecture channels and gathers rainwater. Rather than run-off concentrating in inundated sewer systems, it is available for home use and irrigation. And since oscillations between torrential downpours and droughts are ever more likely in a warming world, technology, including housing, ought to seek out a smooth and supple adaptation to that world through graceful low-tech solutions.

Alongside increased support for local agriculture and state compensation for carbon drawdown, one can imagine a hugely revitalized countryside where there is a great deal of useful and non-alienated work to do. The other side of the coin is peppering the countryside with cultural and infrastructural projects, from effective mass public transit to high-speed internet to health clinics. Free housing, single-unit or communal, could be built disproportionately in smaller rather than larger cities, using local materials. The small US towns, hamlets, villages, and exurbs, which are now ground zero for a murderous white opioid epidemic, could be the basis for a productive renaissance. Poorer Black populations, often scattered in smaller metropolises nationwide, could benefit from reparations payments which take the form of state-built housing, or the state paying the cost for houses people can design or commission themselves.[20] An important experiment in that respect is the Ewing Street Eco-Village Coop Pilot Project, where a variety of home designs are being considered as part and parcel of the larger program for Black national-popular economic renaissance in Jackson, and where designs come bundled with solar panels, passive solar heating for winter and wind-based cooling for summer.[21]

PLANNING AND CONSTRUCTION FOR AN ECO-SOCIALIST FUTURE

Such building could be part and parcel of the revitalization of the public institutions that make up a more decentralized urban fabric. Museums, hospitals and health clinics, and public libraries, alongside more schools need not be concentrated in mega-cities. They could also be loosely dispersed into smaller cities and towns, of 15,000, 30,000, 50,000, 100,000, or 200,000 people.[22]

To the extent that such dispersal of the accoutrements of urbanization in turn entails new building, such new buildings must be carbon negative. And by linking jobs that effectively democratize consumption to the construction and upkeep of such communities, the economy is democratized, and democracy is brought into the planning process. This is not a call to empty out cities, but to remold the smaller cities that nestle neatly into the planet of fields so that they become more alluring to people. Constructing and reconstructing housing, offices, schools, and other public edifices is at the core of a People's GND in the North. Who builds things, how they are paid, what the codes and requirements for new construction are, and where it is built, will lay down the foundations for the next century.

Some demands can be local and negative, accomplished with new zoning laws: Tall buildings have to stop being built. New York and Dubai's glass-and-steel towers are energy sumps and cost an immense amount to cool and heat. Their density is actually wasteful. It is hard to work against gravity and swaying, requiring elevators and concrete and steel foundations.[23] Density can occur on a human scale.

Currently, running and heating housing and other physical plant uses around half of US energy consumption on an annual basis, and around 40 percent of overall emissions.[24] Retrofitting existing buildings for passive solar heating, improved insulation, and other energy-reducing or energy-eliminating features is central. More importantly, that means building with rather than against nature. Such shifts mean retrofitting in the US, and new – but really old – ways of thinking about architecture and design in the Third World. Living roofs and facades, for example, meld the urban with the natural landscape, and can sharply decrease cooling costs.

New construction needs to be built using local materials, which would vary in lockstep with the variety of environments within which people live. In fact, the how, who, and where of construction is a place where the red of labor can meet the green of holistic ecological planning, while creating a mélange of both to increase people's power and to put a stop to local exploitation and break international chains of environmentally uneven exchange.

Local materials are by their nature far less carbon-intensive than trucked or otherwise transported materials, especially because they are extremely heavy. Wood, bamboo, and rammed earth construction all use much less energy than the extremely energy-intensive steel and concrete industries. They have other merits, too. Guadua bamboo in Colombia, for example, is highly earthquake resistant. It can be stronger than steel or concrete. It is a grass and grows almost explosively: up to a meter a week. And as it is grown, it absorbs CO_2 from the atmosphere.[25] When wood replaces concrete in home frame construction, and when the wood waste is used as a biofuel when the house reaches the limits of its lifespan, it makes construction net carbon negative. In integrating construction with landscape management, two fundamental features of these building materials are important. The first is that they are culled from natural processes and husbandry: bamboo and wood are harvested and then they regrow in the same place, leading to net negative carbon emissions, but also requiring the active management of the lands from whence those materials come. Hence, municipal, state, governorate, or federal mandates to use local materials would promote decentralized human populations, since someone would need to husband the land from whence materials come, people would need to extract it, and people would need to know how to manage it in perpetuity to replace existing housing stock as materials wear out. By linking material production to local materials use, construction is woven into the landscape management strategy, through local ecological management and towards global CO_2 sequestration.

Such changes in construction patterns would not be coercive. People adore wood and bamboo homes and masonry, and they can be aesthetically pleasing, beautiful and comfortable places to live and work: the wheaten basket-weave Luum Temple, a bamboo pavilion in Tulum, Mexico, for one. For another, rammed-earth construction has produced stunning alabaster geodesic domes, as the ShamsArd design studio is doing in Palestine.[26]

Furthermore, vernacular building practices and retrofitting practices can be effectively decentralized, particularly when state-level planning encourages such practices. Experiments in milling bamboo, for example, have led to a prototype machine the size of an old microwave, which can cut bamboo into any shape desired. The vision is a field-usable and self-contained box that could design, fabricate, and build on site: "Whereas most digital fabrication technology is high cost and requires economies of scale, this project seeks to develop an affordable DIY machine that can leverage technology to make use of the irregularity of bamboo," says Katie MacDonald, an architecture professor researching bamboo at the University of Tennessee Knoxville.[27] All of this implies relatively more localized if not self-reliant economies at a much smaller scale, with various interwoven sectors. Furthermore, it implies a decentralized distribution of extremely skilled labor, whether foresters, harvesters, or home builders. One example is the recent resurgence of wood as "mass[ive] timber": clumping together chunks of softwoods like pine or spruce or sometimes birch or beech to form larger blocks. Cross-laminated timber made of boards glued together can form huge slabs, up to 90 feet wide, 18 feet long, and a foot thick, and which can replace concrete and steel in building construction.[28] Biochar in clay plasters, hemp building blocks, and straw bale buildings are other carbon dioxide negative options.[29]

As homes and public buildings become artworks, vernacular or otherwise, we open the door to more local and circular economies. Walking through one door brings us to another: a return to aspects of the past, but towards a very different future. Masonry, carpentry, iron-working, glass blowing, ceramic-working, weaving, all the necessary skills for artisanal-style home construction and decoration generally rest on the transformation of local or agriculturally produced materials (iron and steel are obvious exceptions). Even under capitalism, such work is often extremely highly compensated. The problem is not that people do not desire to do such work, and the problem is not that people do not desire to purchase that work. The problem is that our capitalist social institutions have created a world in which the very best work is far beyond the economic reach of poor people and the types of homes in which the poor live.

A socially just non-capitalist planning period would turn every craftsman into an artist, every home-dweller into an art collector and every home into a piece of art. Such homes and other living spaces would also be extremely durable. And because they are based on natural materials, they would incur net-negative CO_2 emissions – "net" because some industrial processes which produce CO_2 may well endure. But part of controlling industrialization is ensuring in the *longue durée* that carbon sinks outpace sources.

TRANSPORT

Moving things and people around at a decent pace is a major part of our modern social systems. Transport enfolds multiple realms of human life. It enables complex human life and human linkage across towns, cities, regions, and continents. It moves people and objects. It is a major sector of employment. Warehousing and transport involve around 14 million people in the US. It is also the leading US emissions sector: around 29 percent of total US emissions, and 20 percent of global emissions.[30] As with all emissions, these are tightly bound to national-class position: around 10 percent of the world's population accounts for 80 percent of total motorized passenger kilometers.[31]

Eco-modernism imagines electric cars can swap in one-for-one for internal combustion vehicles, and air travel can be easily decarbonized using biofuels. While "supply" of raw resources is usually not limited, it does often face increasing difficulty in terms of the energy required to extract it. Such increases in energy required for extraction decrease the efficiency of those processes, in ways which resemble our earlier discussion of renewable energy: it can end up being possible to build a society which looks like a Rube Goldberg device designed around maximizing energy use or car use itself (an alien cosmonaut examining our funky orb would wonder if cars had colonized us. If she were an economic anthropologist, she might discover that in fact the car companies have done so). A recent letter written by a collective of earth scientists noted, for example, that to meet UK electric car targets in 30 years, the islands would need twice current world annual production of cobalt. It would swallow up current world production of neodymium, gobble three-fourths of world lithium production and at least half the world's copper. If mileage did not shrink, to charge that monster fleet an additional 20 percent of current UK-generated electricity would be needed.[32]

Furthermore, "supply" is not summoned up by pristine economic processes but deluged in blood and baptized by fire. Cobalt, one necessary mineral, is kept cheap by a half century of neo-colonial massacre in the Congo, and lithium extraction turns on the mangling of Latin American water tables. Even now Microsoft, Tesla, and Dell are being sued for being party to child labor in the Congo mines that supply material for the batteries that keep their doodads cheap and briskly selling.[33]

However, there are other options, what the philosopher of technology Ivan Illich called convivial technologies, which allow "autonomous and creative intercourse among persons, and the intercourse of persons with their environment," and which would lessen the impact of industrial processes on the environment.[34]

A very large portion of EU, US, and East Asian imports and exports would not occur under eco-socialist planning. Petroleum is by far the heaviest traded good on international markets. Ore is second, and foods of various kinds, especially cereals, and lumber are important. Industrial goods are imported and exported. For the US, the picture is similar: cereal grains, ore and various mineral products, gravel, and hydrocarbons of various kinds absolutely dominate freight by weight. By value, electronics and industrial goods dominate, again alongside the pricey and dense black gold which fuels global development.[35] Since food sovereignty is a key plank of an eco-socialist GND, food trade would be minimized. Since hydrocarbons would no longer be burnt, trade in them would be essentially eliminated. And gravel, for example, ought to be replaced as much as possible with local materials. We would very likely transport far less ore than is currently the case, since ore and metals would be used only when renewable and local materials are not available. This does not mean rejecting goods transport on a large scale and a socially complex, modern, and interwoven world. It could mean, however, that cargo transport for most goods would occur with available renewable sources, whether wind or solar, powering ships and trains. There are already Swedish plans for a wind-powered cargo ship which can traverse the Atlantic in twelve days, only twice as long as current fossil-fuel-powered journeys, or to use wind, solar, and skysails.[36] That would mean long-distance shipping would have to be undertaken more deliberately than it is now.

Even more importantly, intra-urban transportation would occur on bicycles, electronic bicycles, and mass transit, from trolleys to trains, where possible. Private cars would be most often reserved for ambulances and emergency transport – the times and places where society can collectively decide to use the fruits of inherently damaging industrial production in order to protect and convenience human life. Many cities are taking steps to reroute and restrict cars from urban cores via congestion pricing or banning automobiles' entry. In convivial planning, more work would be centered closer to people's homes. Commuting distance is a primary indicator for happiness at work. Under a rationally planned and egalitarian social system there is no reason for people to work outside their neighborhoods, outside of extreme specialization, e.g. physics professors or historians or archivists, those working at necessarily specialized institutions. So planning and transport are intermingled. Planning and urban landscapes should be contoured so as to reduce to a minimum energy- and time-intensive commuting. Studies show that swapping in walking and cycling for short car trips would cut US domestic consumption up to 35 percent and make people healthier, thus reducing healthcare costs: urban planning as a public health measure.[37]

A GND in North and South should invest massively in easily electrified forms of public transports: buses, trams, trolleys, monorails, and subways within cities. They ought to also invest in national and international railroads between cities and between countries, alongside tremendous investments in high-speed rail. And there should be investments in dirigibles and solar sails and other ways to replace flying with zero-carbon aerial or oceanic transport, all decommodified and freely accessible to people by virtue of birthright and managed democratically at the smallest possible scale. Aviation will need to be scaled down while protecting jobs and ensuring a just transition for all of the workers in the industry. Such a transition is easier if access to health care, food, and housing become social rights.

For the countries of the South, such investments could leapfrog over the "car-intensive" modernization model which prevails in the North and increasingly has spread to the South, instead investing public wealth and even more so, climate reparations from the North, into the transportation systems which are the knitting which weaves humans into economically complex forms of social life.

INDUSTRY AND MANUFACTURING: WHO MAKES THINGS?

A central goal of most GNDs is using the state as "master planner" Keynesian-style in order to control *what* is produced, *how* it is produced, and the quantity and quality of jobs. In fact, this has been omnipresent in the GNDs because they are the outgrowth of earlier programs for job guarantees. A just global GND would imply a globally just US GND.

A People's GND would imply a series of major changes in production. First, important re-localization of production, since it is only by dint of unequal global power relations that US consumers can make claims on what is produced elsewhere in ways that global and poorer populations cannot make claims on what is (decreasingly) produced in the US. In Victoria's Latrobe Valley, the Earthworker Energy Manufacturing Cooperative is working on solar hot water heaters to shift from fossil fuels to renewables, while the Redgum Cleaning Cooperative, Earthworker's second worker-cooperative, offers residents of Melbourne green-cleaning services.

Second, Fabrication labs, or decentralized manufacturing hubs of various kinds, along the lines of the bamboo box, allow for sharing knowledge and tools to decentralize production, so long as that production is able to access local material – which is why husbandry practices are so central.

Third, shifting from industrialization, or transformations of abiotic material like steel, and back towards manufacturing, working on living material like wood and cotton. We don't really think about it, but as Colin

Duncan notes, "Perfectly cheap and profitable projects are refused a hearing in our type of modern economy if they are very long-term." Lake Ontario is rimmed with nuclear plants instead of cheap but slowly growing oak forests. In turn, Ontario's universities and schools c. the mid-1990s had mostly steel furniture, which uses a huge amount of energy to make. "Inattention to long-term projects," he adds, "makes our lives now, and later, simply more expensive than they need be."[38] Within the construction industry, a major consumer of manufactured inputs, studies show that shifting from brick cladding, vinyl windows, asphalt shingles, and fiberglass insulation to a wood-frame house which substitutes those other products with cedar shingles and siding, wood windows, and insulation made from cellulose can convert such a home into a net absorber of CO_2 emissions provided that when the house is demolished, the material is recycled instead of entering a landfill.[39] Such a shift has "above" and "below" aspects. Local social movements, municipalities, and ordinances could mandate different kinds of materials use for what is done locally. At the same time, as the movements demand and achieve declining quotas on hydrocarbon burning, state-mandated bans on pollution, and gradually strengthening Third World state capacity, this would create incentives "from above" to shift away from materials whose "cheapness" rests on a right to pollute which any GND would revoke. The cheapness of all kinds of hydrocarbon-based materials, from concrete to plastics, rests precisely on those other cheap or unpaid costs.

Fourth, GNDs need to impose bans on planned obsolescence and shift instead to planned longevity as a collectively imposed standard for manufacturing, alongside easy capacity to repair all kinds of products. Currently, capitalist production makes things which break quickly and are unrepairable or only fixable at great cost or within the vertical service chains of the monopolies themselves – think of your Apple Genius Bar. Capitalism as an irrational and entropic social order has no interest in people being able to repair their own things or a decentralized network of technicians able to do so with their own skills, materials, and tools. A People's Green New Deal would here be in line with the social logic of mass movements. Planning and production are the class struggle. Every change ought to decentralize production where possible and rip away at the socio-physical processes through which monopolies accumulate power and profit. Accordingly, a central plank would be imposing right-to-repair legislation locally and nationally. A People's GND could also impose moratoria on planned obsolescence, or rework industrial manufacturing schematics and materials so that stuff lasts a long time. Once again, it is necessary to note that the easy availability and decisions to discard rather than repair devices like iPhones stems from engineered prices: for minerals, for labor, for the ecological damage which results from extraction, refinement, manufacturing, and export, both directly and indirectly via CO_2 emissions. It is for those reasons that people in the core regularly get new iPhones. If people in the core had to do the labor that *builds* iPhones we would either need to find sustainable ways to automate entire chunks of the production process (not likely) or we would make them as durable as possible, and as repairable as possible – precisely the opposite of their current traits. Empowering local labor outside monopoly extraction would be easier once repair and fixing is radically decentralized and democratized.

Fifth, there is an aesthetic question, with a relationship to mass production and needed consumer commodities which should be the object of critical exploration and conditional embrace. In many ways the planet is over-industrialized. At the same time, it is striking that objects considered luxuries, only available to those with more social power, are often objects which come from time, skill, love, artisanal attention, and in one way or another emerge from the land. Handmade leather goods rather than mass-produced plastic jackets and bags. Wooden and glass rather than plastic furniture. High-end cashmere and cotton rather than polyester and acrylic. Goat's cheese rather than Polly-O. Artisanal bread rather than Wonder Bread. Those with class power prefer consumer goods from smaller-scale manufacturing rather than industrial sectors, keeping in mind the blurry line between industry as working on abiotic, or dead, matter like ore while manufacturing works on biotic matter: cotton, linen, wood (it is blurry because one can make cotton textiles, to take one example, in decentralized and less industrialized, or highly centralized and industrialized ways). Mass-produced industrial tchotchkes are relegated to the poor people of the core. To get much more of what we currently get by industrial processes through artisanry, smaller-scale and decentralized manufacturing, and judicious land-tending does not imply a reversion in anyone's quality of life. It would be an anti-capitalist qualitative advance: communal low-tech luxury.

A sixth shift is the temporal and organizational logic of industrial manufacturing. Questions of energy use, labor, manufacturing, and industrialization merge here. If eco-socialist planning relies in a larger measure on understanding that current patterns of production are structured based on a certain kind of constantly available energy, a more supple approach to manufacturing might solve a large portion of the intermittency problem. Globally, industry uses around half of end use energy. Some of these processes rely on purely mechanical energy: turning, polishing, milling, hammering, crushing, sawing, and cutting. These could be run with intermittent power, as can great parts of food production like olive pressing, or hulling and grinding grains, rock and ore crushing, or textile work, such as preparing fibers, weaving, and knitting. As Kris Decker notes, "intermittent energy input does not affect the quality of the production process, only the production speed," and, "Running these processes on variable power sources has become a lot easier than it was in earlier times. For one thing, wind power plants are now completely automated, while the traditional windmill required constant attention." Factories could run on a mix of wind and solar. Such shifts do not mean reducing production or consumption. They simply mean fitting industry and manufacturing into the vicissitudes of non-hydrocarbon energy sources. Producers could make items in seasons or moments of abundant energy, and store them "close to consumers for sale during low energy periods," effectively turning clothing or furniture or cotton into "energy storage."[40]

CONCLUSION

A People's Green New Deal is about building eco-socialism. Building a new world does not happen overnight, but to build a new world one must have an idea of the world one wishes to see and take steps to get to that world. There is no iron law which states people need to work a certain number of hours per week or that they should be paid far below the value of what they produce. Demands for living wages on weekly or annual bases – say something like $40–100,000 a year – could dovetail with demands for vastly decreased working hours. In this way, workers would immediately acquire a huge portion of historical productivity increases, and further productivity increases would go to decreased working hours and climate debt settlement. The value reacquired from capital would turn into carbon-free free time. Such massive minimum wages would be and always have been the fruit of political demands and social struggle. There is no reason such demands are unreasonable.

Comprehensive social rights, especially housing and healthcare, the largest budget items for working families, are also necessary. Affordable housing – with prices capped at 15 percent of salaries – could be the child of severe rental control laws, housing cooperatives as the property of unions, or government vouchers for housing. Real estate speculation and real estate investment would disappear. Furthermore, housing could be built locally, owned by the state or city or town, and contracted out to living-wage-paying cooperatives using locally sourced and renewable materials. Governments can set their own standards for securing materials, and can include environmental costs and consequences.

Similarly, healthcare should be more or less free or available at token cost. It could be transformed into an environmentally soft-footed, knowledge-intensive system reminiscent of Cuba or India's Kerala. Transport would also be decommodified locally, and paid for via wealth and real estate taxes, set fairly low – $500,000, or more in metropolises with elevated housing costs – so that the wealthiest would pay the costs of universal free access to transit. Longer-distance travel could be similarly decommodified, although probably it would have to be rationed – which it already is by the market. With manufacturing and construction work paid a high wage, more people would be interested in such work and more value would circulate locally, creating wider local internal markets, and creating ever more socially useful and carbon-free or carbon-negative work locally. People would be paid for the work of caretaking and teaching and rearing children at the same rates as other skilled labor, further enhancing the power of labor against capital and carbon-free work against carbon-intensive work.

Many of these demands are actually basic core social democratic demands, although we need to pay more attention to the technologies which go alongside redistributions of social power. They clearly go far beyond the current elected progressive policy platforms. But we should not redefine socialism or eco-socialism by identifying it with the left wing of the possible. Rather, we should sharply distinguish socialism from political programs which seek to confront some sectors of capital, contain others, and preserve still others in an attempt to arrive at an entente of sorts with the latter. None of the more radical demands are objectively impossible. They are only difficult, because social and political power is in the hands of wealth-holders.

Where a People's Green New Deal substantially differs from the social democratic green programs is when it comes to land planning, agriculture, and the relationship to the Third World through climate debt and the national question. The next two chapters deal with those two sets of topics.

6

A Planet of Fields

Imagine a world of cities with mottled yards full of chestnut, pecan, or breadfruit trees, below them perennial raspberry bushes or the dusky crimson and pale yellow of pomegranates, trellises wrapped in grapes and squash, and lower, lifting out from healthy obsidian soil, a palette of landrace tomatoes and peppers. Roofs are coated with patterned raised beds. Side-yards feature fishponds, which receive all kinds of manure, fattening up their inhabitants. Greenbelts surround all cities, as popular planning converts sprawling suburbs into farmed swards. High-speed trains link bigger cities to smaller ones, smaller ones to capillary lines leading to outlying hamlets kitted out with high-speed internet. Ranchers manage grasslands roamed by a bouquet of native breed animals that churn cellulosic matter into the soil, producing negative-CO_2 milk and meat. Intercalated forest-fields overproduce walnut, fodder, and myriad maize and wheat varieties. In some places, the burgeoning low-tech of perennial cereals inject root structures far deeper than people are tall into prairies. There is a lot to do beyond farming: Skilled technicians live in the countryside and smaller villages and estates, and manage high-voltage smart grids, local renewable storage systems, and decentralized windmills, while artisans and decentralized manufacturing processes local agricultural materials, supplementing and replacing – where it makes sense – large-scale mass-market industrial goods.

How to get there? Land Back restores Indigenous treaty rights and decolonizes the continent. Land-to-the-tiller agrarian reforms shatter large capitalist plots into smaller ones workable by non-patriarchal familial units or organized in cooperatives. This is done with due attention to the history, especially, of Black land loss. Prices change. They are fixed by political decisions that reflect popular needs. Nations set parity prices for their agricultural goods, so that prices reflect the labor needed to do ecologically restorative farming. In turn, within countries, agriculture and manufacturing and services – to that extent they would remain separate and specialized tasks – would exchange using prices based on labor inputs, including the labor needed to protect the environment.

Living wages alongside a gift bag of social democratic rights for former itinerant farmworkers ensure jobs are good everywhere, so people do the work they wish to do. And if not enough people want to do land-based work, they would be paid more, reflecting the infinite social utility of protecting the environment through land management, including that involved in farming. Land restoration, agroecology, and pasturing are the social labor which ensures CO_2 absorption, saving the planet and safeguarding the future, and is cherished. Food is excellent and uniformly ecologically farmed, a form of preventative medicine which prevents diet-based pathologies. Because food is also generally more locally procured, people see the value of producing good food and the impacts such production has on the landscape. No deforestation. No catastrophic soil losses. No mercenary economists summoning up Malthus. No demographic chart-mongering about starvation. No more zoonotic disease reservoirs from concentrated animal feeding operations (CAFOs), clear cuts, and genetic doppelgangers multiplying amongst animal populations from the sorcerer's apprentice of industrialized animal agriculture. And the real magic is not magic at all: atmospheric CO_{2e} descends in notches, each one a small miracle, to its late-industrial safe zone of 300 parts per million.

Those shifts would be revolutionary. And like nearly all of the revolutions that ripped apart the capitalist-colonial chains of the twentieth century, the one yet to come depends on socialist agriculture. Because poor people would control their lives and labor and environments and lands, they can staunch the ecological and value hemorrhages of unequal exchange, while in the core, part of the burden of transformation is shifting to a system of land management which is popular, socialist, just, and makes pulling CO_2 from the atmosphere into the land part of the labor of a socialist civilization.

Planning for a Planet of Fields means putting more agriculture in cities, a total shift to agroecology, making agriculture more pleasant and rural areas a little more urban, encouraging cities to have a sustainable relationship with their hinterlands, reorienting national-level planning and laws of value to ensure the priority of humanity is locking CO_2 into the Earth, and using agriculture to produce food as use-values for people rather than exchange-values to enter the maw of Western accumulation. It means drawing on the best of the past, not as curio, retrograde traditionalism, or out of nostalgia, but to find the way to a better future.

Such a mission requires a lot. First up is dispelling or simply transcending calcified anti-rural prejudice, the millstone inherited by Western Marxist and liberal thought from the colonial worldview in which it was born and bred. What if utopia was a little less urban and a little more rural – more like a Planet of Fields?[1] This should only be jarring to some. For others, it would not shock at all. In the words of anthropologist Kyle Powys Whyte, "food and medicine" were always "among the most heavily covered topics in relation to climate change" for the Indigenous.[2] Agriculture is important because food is important, one of "those recognized elements of a 'good life' that are most strongly cross-cultural," bound to people's identities, a way people cherish time together.[3] This universality has made food sovereignty an important radical banner in the South and North.[4] As the banner of Land Back is flown more and more, perhaps it is time to listen more attentively, not as posture, but to actually learn and take political and ecological guidance, from those who are holding that banner high: the victims of the West's visions.

TRADITIONAL AGRICULTURE: YIELDS

Before describing our current capitalist-imperialist food system, it is good to remember just how different it is from the pre-colonial world. Before the colonial conquest, much of the US was a managed forest-garden. Through controlled burns, bison runs, terracing, earth works, and farming, the entire continent ranged in between what used to be understood as hunter-gathering and settled agriculture. It was not a pristine "first nature," wilderness, or unsettled.[5] People lived in and remade nature.

In the Western areas, consumption was astonishingly varied: "the sweet, thirst-quenching taste of manzanita cider, the puckery sour taste of sourberries, the pungent taste of saltgrass, the salmony taste of bracket fungi, the spiciness of tar-weed seeds, and the slightly bitter taste of Mormon tea bread."[6] Yields per unit of labor, per unit of land, and per unit of energy were impressive. Indigenous farming before the settler-colonial invasion in the eastern United States returned an estimated 5–15 calories per calorie input.[7] Further west, managing the landscape included methane-belching bison, to the tune of some tens of millions alive at any given time. Production of the Three Sisters involved intercropped maize, beans, and squash on mounds. The fields were not plowed, decreasing oxidation of soil organic matter and soil erosion and boosting fertility. The combined yields of the three plants increased through complementarities and synergies, yielding more than if each were a monocrop: 12.2 million calories and 349 kilograms of protein per hectare.[8] (By comparison, contemporary corn production yields around 30 million calories per hectare; not nearly all of which is used for humans.) Such mixes provided myriad nutritional benefits for their eaters. In Hawaii, taro, sweet potato, and stone fishponds provided enough food for close to a million people.[9] As a consequence, "Native North Americans were among the healthiest in the world."[10]

THE POLITICAL ECONOMY OF INDUSTRIAL AGRICULTURE

The modern agriculture and food system is the fruit of settler-colonial advance, capitalist primitive accumulation, and accumulation on a world scale, leading to ecocidal landscape destruction and devastating health outcomes. From the outset, the huge plains of Australia and the Americas, including Canada, the world's breadbaskets, came into being as a result of colonial genocide inflicted upon Indigenous peoples.[11]

The US food system, and to a lesser extent, the world food system, is oriented to production not for people, but for profit. On a world scale, 84 percent of farms are smaller than two hectares, but they only operate around 12 percent of farmland. The largest 1 percent of global farms operate over 70 percent of global farmland. And the trends North and South are worsening, not improving: in the US, the largest 7 percent of farms produce 80 percent of production value. In the EU, fewer than 3 percent of farms cover over 50 percent of the farmed land. In agricultural Tanzania, 108 recently imposed large farm investments control more land than the smallest two million total farm entities.[12] The foundation of capitalist agriculture is private and highly concentrated ownership of the most basic element in production: the land.

Despite all that, on the one hand, on the global level, at least 50 percent of food, maybe more, is grown on smaller family farms, using various amounts of capital-intensive inputs.[13] Many of these farms have at least a foot in agroecology.[14] The remainder is grown by massive agribusiness corporations, or far larger farms which sell directly to those corporations. These farms are ever growing, gobbling up neighboring land and leading to ever more marked property concentration worldwide. And they use highly capital-intensive methods. Their aim is to produce as cheaply as possible in terms of dollar prices. At the same time, they are woven into a global corporate system. Seed, fertilizers, pesticide, farm equipment, and irrigation-technology companies are highly concentrated. Those firms want as many farmers as possible to buy and use and rely upon their technology. The buyers of agricultural commodities – soy, wheat, corn – and those companies which supply supermarkets want to buy farm products as cheaply as possible. None of those companies are interested in farmers or farmworkers earning a living wage or ensuring that the food people eat comes from healthy crops.

Furthermore, because margins are often very low, the companies need to buy and sell a lot of crops to make money. In this way, as capitalism enters and reshapes agricultural production, larger and smaller farmers alike produce and overproduce, in part because they are often trying to get out of debt. Government subsidies for cereals also lead to overproduction. Such cereals – corn, for example – are often processed into highly unhealthy corn syrup. Or soy into soy oil. Overproduction of wheat and soy led to US dumping of wheat and soy oil onto the Third World, from the 1950s onwards. In countries like Colombia, Tunisia, and Egypt, this policy damaged smaller farms, held land-to-the-tiller agrarian reform at bay, and destroyed national capacities to feed themselves, while Africa from the 1970s onwards became a structural food importer.[15] In the Third World, most countries opted for focusing on agro-export. They emplaced Green Revolutions in their cereal sectors from the late 1960s onwards, using input-intensive methods, rather than trying to develop in different ways using their own capacities. This northern-assisted decision, too, was a great boon to the huge northern conglomerates which cornered the supply of those inputs. In the 1980s and the 1990s, this process accelerated. Southern agricultures, under pressure of the international financial institutions, opened up even more as a result of structural adjustment policies. Countries stopped supporting even medium-sized farmers and dismantled strategic grain reserves, while the Western-celebrated collapse of the Soviet bloc led to a collapse in world grain demand, helping keep prices lower at the costs of shattered lives in the former Second World.[16] In 1995, the World Trade Organization began to promote "intellectual property rights" worldwide for genetically modified soybeans and maize, making southern agricultures even more dependent on northern inputs. In the process, major monopolies such as Monsanto, Bayer, and Syngenta, as well as Cargill, Coca-Cola, ArcherDanielsMidland, Tesco, Walmart, and Carrefour compounded their power over the world food system.[17] They are, of course, sited in the North.

Finally, there is the imperial division of labor. During the colonial period, the Third World produced commodities like spices or coffee which could never be grown under any conditions in the First World. They were then drained from the Third World, producing widespread famine.[18] That process slowed but did not stop with decolonization – which raises the centrality of national liberation to a reformed food system. With the advent of neoliberalism, dollar-cheap but land-and-labor-intensive tropical food imports, such as out of season fruits, vegetables, and other tropical goods, have been grown ever more on Third World lands.[19] These crops simply cannot be produced in the North (except in small quantities in greenhouses). They rely on ultra-cheapened labor in the periphery and represent claims on the land and water and lives of those countries. Part of how that labor is made cheap is through vast labor reserves, which are maintained, and land cordoned off for export crops, through ruinously unequal rural agrarian structures, widespread malnourishment, and of course, punishment and siege against countries which defy the capitalist or colonial agrarian structure.[20] The prerequisite for such a system of feeding has been preventing agrarian reform. As we will see, changing the food and agriculture system means changing who owns land.

The ecological and social consequences of these policies have been enormous. Modern agriculture uses artificial and resource- and energy-intensive fertilizers, which leach into water tables, poisoning groundwater. They produce massive run-off, inciting algal blooms at riverine termini. Modern agriculture compacts or over-tills the soil, causing erosion or other forms of soil damage. The land can no longer absorb water, in contrast to healthy soil, which is produced by mixes of animals and poly-crops, cover crops, and no-till farming. The effects in the US are not hypothetical: the massive 2019 floods that inundated the Midwest would largely have been avoided if the soil had been healthier. Whereas traditional agriculture suppresses pests through promoting predator-healthy habitats, industrialized agriculture soaks fields and their watersheds in pesticides, leading to bird and insect apocalypse, and infiltrating human tissue.[21] Losing insects leads to declines in pollination and nutrient cycling. As insects die, so do the creatures that rely on them for food, including birds which play a vital role in seed distribution.

Furthermore, agriculture's energetic basis is now topsy-turvy. So-called "traditional" agriculture relies on plants' capacity to efficiently gather solar energy and convert it into forms usable by humans, an organic machine of absolute brilliance and grace. The new energetic basis of agriculture turns that logic on its head, using past flows of heat and light from the sun in the form of fossil fuels: "a change from 'using sun and water to grow peanuts' to 'using petroleum to manufacture peanut butter.'"[22] Corn in the US in 1970 produced just 2.6 calories per calorie invested.[23] People estimate that now, advanced industrial societies use 4 and 15 calories for each unit of food they produce (interestingly, as little as 8 percent from mechanization).[24] Such energetic inefficiency is only very partially due to industrialization. Of the energy used in the food system, a third is for production, a third for processing and packaging, and a third for distribution and preparation. Importantly, such interlinked chains do not manacle peoples and environments nearly as much in the "underdeveloped" countries.[25]

The North is "adept" at growing cereals, provided we do not mind topsoil blowing and silting up rivers, nitrogen fertilizers inciting hypoxic dead-zones in the Gulf of Mexico, and the specter of silenced springs as biodiversity evaporates in such factories-in-the-field. Such farming skill is one that treats living soil as dead. When we do so, we ought not to be surprised when the graveyards spread.

Furthermore, the quality of much of the food we eat lacks taste and nutrients, blights that worsen along lines of class and race: the poor with highly processed fast food (served to them by other segments of the poor) or other food soaked and laden in trans fats, doused in sugars, and otherwise ludicrously unhealthy even if affordable in dollar terms. It is furthermore overwhelmingly the poor who are forced to eat the products of the US agro-industrial food system: grain-stuffed and hormone-pumped animals versus the sustainable and grass-fed steers which are better for the environment but cost far more than poorer consumers can afford.

Modern food chains are also wasteful, substantially because of capitalist overproduction: if we produce more than we need, much of it will go to waste, whether (over)processed into ethanol or nutrient-free spongy bread, or fruits and vegetables discarded for not meeting aesthetic standards of one kind or another.[26] In the US, around 6,000 calories of food are produced per person per day; 30–40 percent of it is lost, including in the many stages of capitalist production and transport, and around 10 percent goes to animal feed. This leaves around 3,700 calories per person. Humans need far less than that, although perhaps a more active and nature-engaged population would be a hungrier one, too. On a global basis, total calories available per person are closer to 3,100, with less food waste, especially in the Third

World, including China and Greece. Food waste is intimately related to the syndrome of production that links industrial monocrops to urban and slum consumers. Elongated production chains are blighted with weak links and rust. They only make sense from the perspective of the monopolies that forge them and use them to strangle the planet and its poor. In rural areas, food waste is far lower. The longer the commodity loops are, the more food is wasted, the more such waste cannot return to the soil to restore its fertility, and the more carbon dioxide is produced when shipping and flying such foods around the world. City planning plays its part in inducing more food waste, while also removing that waste from the nutrient cycle.

Industrial agriculture has also meant farms without people. From 1940 to 1970, the push-pull factors of oil replacing rural labor, the seductions of suburban modernity, and evermore onerous difficulties of rural life halved the number of US farms. They went from about six to three million, and from 23 percent of the US population to 4.7 percent by 1970. By 1995, it was 2.2 million and 1.8 percent. Along racial lines, de-agrarianization was even sharper – Black farmers were 14.3 percent of farm operators in 1920 and 1 percent in 2000.[27] Farmworkers themselves are predominantly Latino – a workforce partially made by the attack on the Latin American countryside through free trade agreements.[28] Additionally, while the US has literally de-agrarianized, if one counts the number of workers elsewhere occupied in the food chain – in slaughterhouses mechanically ripping at chicken carcasses until their hands are quivering with carpal tunnel syndrome, as vividly recounted in the work of feminist geographer Carrie Freshour – or in the fast food industry, or for that matter at Walmart, the US's largest employer, then the proportion of the US population working in the food industry overall is not all that much lower than the percentage of some Third World countries working in the agriculture sector.[29] Labor in the food process has been displaced by machines, but labor is still quite present in the food system, and once again, it is very tough labor. "Modernization" has hardly meant the bettering of peoples' lives.

Finally, "modernization" of agriculture has not saved people from hunger. Worldwide in 2019, two billion people did not have regular access to nutritious, safe, and sufficient food. Overwhelmingly, those people are concentrated in the Third World, including in countries which have seen a great deal of agricultural "modernization," like India's Green Revolution.[30] Fourteen percent of India and the Philippines are undernourished.[31] In Yemen, under a US-planned assault, 16 million people are food insecure, a telling example of why "the national question," or freedom from foreign intervention, is still the foundation of popular development.[32] Meanwhile, the capitalist organization of food supply doesn't just produce problems of quantity, but also quality. As the Food and Agriculture Organization reports, "overweight increases in lower-middle-income countries are mainly due to very rapid changes in food systems, particularly the availability of cheap, highly processed food and sugar-sweetened beverages."[33]

STEPS TO PUT AGROECOLOGY AT THE CENTER OF PLANNING

Agriculture must be at the center of popular planning if a People's GND in the core is to be internationalist and eco-socialist, and if popular provision is to break with neo-colonial commodity chains. From each according to their ability means the US will have to produce more of what it consumes, without foisting that labor onto an illegalized working class. Wrangling world-looping commodity circuits onto the continental US means that hard work cannot any more be the work of the world's poorest for which they receive the worst salaries. Instead, that labor will have to be either eliminated – the techno-futurist tomfoolery of heaping more energy-gluttonous technology onto a world in which we need to shrink core energy use – or distributed fairly amongst all people in the US.[34]

A new planning system is the basis for such a transition. Shifts will occur partially through command-and-control measures based on shrinking quotas for carbon emissions, and the need for a high degree of organization in order to achieve the progressive socialization of property and planning. If social wealth is generally allocated based on labor inputs, relatively labor-intensive agroecology will receive more compensation. And since radical agrarian reforms will break apart large estates, in periphery and core alike, matching up labor and land is eased considerably. Command measures ought to extend to phase-outs of industrial agriculture, which has no justification for existing, and wholesale shifts in research spending away from conventional agricultural research and towards agroecology.

Farms should stop producing as much as possible, especially in the northern capitalist states. Instead, farms should have quotas based on how much they can produce using permanently sustainable production methods, including crop rotations and conservation plantings which regenerate soil and shield biodiversity. Corporate agribusiness should be simply dismantled through the nationalization of such corporations, and the use of their labs for people-centered agroecological research. Land should be redistributed everywhere. Each country should have total control over the food import and export trade, so that food dumping is impossible. Communities rather than farms should be in control of water, seed, and eventually land.[35] This is the agenda of food sovereignty.[36]

Modern agriculture, including agroecology, really has one tough thing about it: the difficulty or extent of hard manual labor as a part of humanity's common labor. Whether this is truly a trade-off depends on perspective. It will seem onerous from the perspective of the Western anti-peasant and anti-manual labor convention (although people do love going to gym). The need for a higher percentage of humanity to be engaged in agriculture than the US's currently picayune 2 percent may seem not such an apocalyptic shift not merely when one considers the alternatives, but even when one looks to how so many people fill their free time when not under the lash of the wage system: in the core we care for home gardens, food and ornamental. We gather in community gardens and we go berry picking. And we cook, a central element of social reproduction, one which few seriously suggest automatizing, and one that is overwhelmingly the task of women and the lower class.[37] Furthermore, with prices for produce reflecting labor costs, farmers will work a lot less hard than they do today. They will work even less hard as labor-saving, appropriate-scale technologies become more widely available and become the object and beneficiaries of more and more research.

Putting agriculture at the center of planning is not a call for a return to a pre-industrial civilization. It rests on recognizing that capitalists have over-industrialized the planet, blanketing it in a gratuitous technostructure and swamps of metallic and chemical waste. The species-specific things humans need and enjoy, necessary in adequate quantities and in as-good-as-possible quality, are shelter, food, medical care, and the industrial or manufacturing processes which help us access the first three. We also need the means of culture. Most of them, like books, arts, literature, and theater, are not themselves strongly dependent on industry.[38] Aspects of medical care, of course, and computers, and transportation are absolute priorities for industrial processes.

MOVING TO AGROECOLOGY

National-popular control over food and farming is an entry point into restructuring our world. Food is not merely an entry point into social reproduction, the hidden abode of home and hearth. It is also a hatch which when opened up, allows us to see and manage an even larger element of social reproduction: the human relationship with the non-human world. Agriculture is unique among production systems in that it can be smoothly integrated into natural cycles and can offer a warning when one thing or another goes awry. And unlike almost any other productive sector, agriculture requires minimal trade-offs. It doesn't require people to weigh and sift merits of efficiency versus cancer plagues, more and newer tech versus slurry ponds from ore processing. It is not just that agroecology does not have those kinds of trade-offs. It can turn homesteads into sanctuaries for much broader realms of the ecology. Because animals are mostly indifferent to ecologically managed landscapes versus non-yielding forests and grasslands, agroecological production is a sanctuary for biodiversity, and is a "firebreak," in the words of biologist Rob Wallace, against new plagues.[39] And such biodiversity-boosting and water guarding, ceteris paribus, should not affect food availability.[40] Properly tending to the land improves resiliency against floods, mudslides, and drought. Furthermore, agriculture, through harnessing photosynthesis, is the most brilliant and effective way we have to gather solar energy and convert it into a form usable by human society.

Agriculture is also, potentially, highly democratic and decentralized, having a kind of elective affinity with a centrifugal planned eco-socialism. Sustainable farming is knowledge-intensive and self-organized at the level of the producer. Agroecology is the technological keystone of this Great Transition from below. The term refers to applying scientific experimentation to and the formalization of the processes underlying traditional farming systems.[41] From poly-cropping to cover crops, from multiple-storied forest-gardens to raised beds, from Vietnamese Vuon–Ao–Chuong integrated fish-garden-pork plots to the maritime gardens of Gabès, these systems have a number of traits which could combine to be the wattle and daub of a revolutionary transformation in the world agricultural system.[42] The very best kinds of farming reduce work from the amount of toil demanded by run-of-the-mill industrial farming as with no-till techniques. Such "low-tech" agroecological methods rest on truly astounding collective wells of local knowhow. The challenge is finding analogues of such practices that try to reach equivalent labor-energy input-output rations, yet can support far higher population densities, through a transition to a knowledge-intensive agriculture.

Modern agroecological systems have a number of key traits relevant to guarding the environment. One, they recycle biomass. Two, such a style of farming strengthens the "immune system" of the larger farming system by promoting natural enemies of pests. Three, they promote healthy soil, adding soil organic matter. Four, they minimize loss of water, energy, or nutrients by conserving and regenerating soil, water, and biodiversity. Five, they promote species- and genetic-level diversity over time and space. And six, they enhance synergies amongst various ecological and biological processes. Such systems work within rather than against the tendency of natural systems to grow polycultures and recycle wastes. On marginal lands agroecology may also out-yield conventional systems, whether for cereals or agroforestry. They also fail far less frequently than do monocrops in reaction to climate change-induced disasters and better tolerate or bounce back from extreme weather events.[43] Furthermore, agroecology introduces firebreaks that prevent the passage of animal viruses to human populations.[44] Finally, agroecology works against scientific and technological dependence of the South on the North – a form of development by popular protection.

There are a number of immediate social and technical changes which can be made to move away from the current system. To eliminate chemical fertilizers, we should replace them with organic or green manures, or even night soil. This is easier to accomplish in the South. It must also be done in the North to close the nitrogen cycle. That implies ensuring larger human population concentrations can recycle waste rich with nitrogen and phosphate, whether through urban gardening, peri-urban plots, or restructured suburbs or exurbs. Edo-Tokyo did this several hundred years ago. The Chinese collected waste in terracotta jars.[45] Shattered nitrogen cycles would be replaced by composting toilets which would sew shut metabolic rifts. A partial re-localization of agriculture and urban farming can facilitate such ecological closed cycles, as distinct elements interlock into a seamless network.

Furthermore, urban planning ought to be nested into the surrounding ecology: as many parks as possible, gardens instead of lawns on every square inch of the city, green roofs. Urban planning transforms cities. They would still be relatively non-agricultural places. But they would also be elements of what biologists Ivette Perfecto, John Vandermeer, and Angus Wright call Nature's Matrix: a lustrous patchwork wherein high-quality agroecological swathes are intercut with agroforestry and non-farmed land where climax communities can bloom.[46] The urban fabric comes to be threaded with wildlife patches of daub olive and forest green. Streets ought to be lined with trees. Every home and street-lined boulevard, rooftop, façade, brown lots, should become a communist victory garden. Urban agroforestry should fill green spaces currently occupied by that idiocy of post-war American life, the lawn, or purely ornamental trees, community or public parks, rooftops, backyards, and for that matter, most golf courses. Chestnuts, honey locusts, and carobs could replace cereal brought in from elsewhere. Urban farming should be accelerated through agroforestry: chestnuts, walnuts, almonds, and woody perennials in the northern hemisphere, bananas, figs, coconut, and breadfruit in the southern hemisphere. Urban mushroom gardens should be used to process urban waste. Such work would not only be socially and environmentally rational. People would love it, as they already love parks and community gardens.

Annual vegetables like tomatoes or perennials like pear trees would replace goods trucked in from elsewhere – trucking which requires fuel for the trucks and a larger fleet of trucks than would otherwise be the case. Havana's urban gardens, especially the *organoponicos*, provide 70 percent of the city's fruits and vegetables.[47] Urban farming in Tongaat, South Africa increases food security, and is part and parcel of social reproduction for large percentages of the urban population in South Africa, Ghana, and elsewhere on the continent, with a large role for women.[48] Bologna could produce 77 percent of its vegetable needs, super-dense Boston 30 percent.[49] To the extent that food is produced locally, people could apprentice in agriculture, and learn that some portion of each year or decade would be spent on farms – unless farm work turns out to be something enough people adore that there's no need to allocate it fairly based on voluntary-collective decisions. Furthermore, chronically underemployed labor would instead go towards decommodified urban food production.

Cities with more greenery are cooler, lessening electricity costs and population displacement, and leaving energy to be used for other urgent non-replaceable necessities.[50] Such tools would tamp down heat island effects, provide shade and reduce energy costs directly through ambient cooling and indirectly through reduced need for energetically cooled places and spaces. In turn, cities would be at least relatively more self-sufficient, although metropolitan areas would be more so, since some cities are too big to grow much of their own food, and no one wants to forcibly relocate people out of cities.

They would drawdown urban CO_2 emissions through locking carbon dioxide in cellulose. Intercutting and seeding trees, shrubs, wildflowers, and non-chemical-doused polyculture gardens within cities chip away at the austere, hard, industrial "citiness" and turn them more into the types of rich matrices which nurture biodiversity. Green spaces as small as 150m² increase bird diversity within cities. Rooftop gardens provide similar benefits, which is crucial when we consider that roofs are around half the impermeable surface of cities. Green living roofs absorb CO_2 and can sharply reduce buildings' energy use – up to 50 or even 70 percent in southern Mediterranean cities. They also clean up pollution and lighten urban sewage systems' load during deluges, slated to be one the key climactic shifts of global warming.[51] They also reduce the heat island effect and could potentially be sites of food production, and finally could be part of urban "extensions" of nature's matrix if walls and roofs are woven together into a connected green fabric.[52] In Singapore, large and low urban roofs, the calling card of a non-dense city, with large low-slung buildings uninterrupted by roving automobiles, were home to bird and butterfly populations, boosting biodiversity.[53] Overall, making cities greener would not be a simple question of this or that urban garden, but could seek as much as possible to make cities part of the thatch of complex matrices which add to biodiversity, moderate temperatures, and allow for some food to be grown at home – agroecological victory gardens in the North as part of a planned economy and environmental management.

CARBON DIOXIDE DRAWDOWN AND FARMING

Most beguiling for climate change prophylaxis, agriculture and land management can draw CO_2 from the atmosphere, using current technology as part of a holistic drawdown path. Right now, major agricultural emissions come from land use changes, including the destruction of peat and deforestation, and methane and nitrous oxide emissions from fertilizer use, rice, and livestock – bracket, for now, livestock and the myth of methane. Deforestation and the eradication of peat primarily occurs for oil palm plantations in Indonesia or soy plantations in Brazil, sold to northern consumers via agribusiness, or incorporated into production chains as low-cost food items, which maintain the prices of labor relatively low on a world scale (most soy is part of the animal production chain, and a Planet of Fields rejects feeding animals soy). So we could stop many emissions relatively easily.

Trickier but more interesting is absorbing the CO_2 already in the atmosphere. The capacity of degraded soils, deforested terrain, damaged grasslands, and other biomes to absorb excess CO_2 is staggering. The IPCC's estimates for what is possible from natural climate solutions range extremely widely, from less than one $GtCO_{2e}$ of mitigation, or prevention of current CO_2 emissions alongside absorption of CO_2 already in the atmosphere, to over 20.[54] Other estimates put the total for natural carbon solutions as between 20.3–37.4 $GtCO_{2e}$, although not all of that is "cost-effective."[55] For reference, the world currently produces around 36.4 $GtCO_2$ per year, not including land use change.[56] Including land use change, the total is around 50 $GtCO_{2e}$. The vast range of what is possible using the land is the child of our great ignorance about just how successful CO_2 absorption through land management and better farming and pasturing could be, with epistemology having its own political economy – the money has not, until recently, been in natural carbon drawdown.

If research and social planning went fast in this direction, we could possibly reduce total emissions between 21 and 37 percent, and at the very high-end estimates, have negative emissions totaling around one-third of current emissions.[57] If we wish to get back to pre-industrial CO_2 levels, we need to pull a trillion gigatons of CO_2 from the atmosphere, or around 70 years at 15 gigatons a year. At the current state of knowledge, it is unclear what is or is not possible. What is clear is that such measures tend to not involve real trade-offs: instead, they can simultaneously improve environmental and social outcomes. Technically, the possibilities are myriad. Soil is a huge terrestrial storehouse of carbon, and small changes in soil's carbon count could cascade into massive reductions in atmospheric CO_2. Of the 100 best climate remedies on the Drawdown project website, around 85 are linked to permaculture, and they "essentially condense to one: simply saving our soils."[58] Other measures include restoring degraded and abandoned farmland through artisanal replanting, whether of wild grasses or of trees, the labor-intensive work of "rewilding" through introduction of keystone species, whether decimated predators or herbivorous "ecological engineers" who can remake degraded or destroyed territories into shifting pasture-forests, maintained by grazing.[59]

FOREST-GARDENS

One of the first critical steps to using nature-based carbon drawdown and mitigation is forest tending. Ceasing cutting down trees yields a bouquet of benefits. Old-growth forests store far more carbon within their biggest trees than fast-growing genetically engineered monocrop tree plantations and are replete with biodiversity.[60] Management and tending of forests is critical: forests as places where people get wood, non-forest wood products, and other gifts. Leaving forests to become untrammeled old-growth stands is best. But there are other paths: reduced harvest rates and increased rotation times. More importantly and bringing us to the manufacturing stage of the production chain, wood production needs to be made as efficient as possible and wood use should move to longer-lived products. No more plywood or Ikea built-to-break products that shatter when moved an inch. Furthermore, forests should not be used for fuel, which brings us to the synergistic and systems-level approach: conserving forests as a peak and precious CO_2 stowing technology relies on advanced technology and power distribution and demands a certain level of industrialization. It also relies on reskilling and locally transforming wood into furniture through carpentry becoming the dominant means for furniture-making.

Reforestation of previously deforested areas is a keystone of CO_2 drawdown strategies. This need not be done using nature preserves, let alone their hyperbolic expansion into half the earth which in the Third World would brutalize poor consumers and producers alike.[61] Nor should we be planting trees where they have never been, or where some misleading models imagine they were some hundreds of years ago.[62] Instead, reforestation ought to be done in large part through forest-gardens, a valuable element of a polyculture-based development model, and by restoring forests or mixed growth to places that have been clearly shown to have been deforested. Social development and environment protection should go hand in hand. To take a southern example, in the arbors of Sumatra, former industrial workers have set up land occupations, planting "diverse agroforests of high-value fruit, spice, and hardwood smallholder commodity tree crops."[63] Elsewhere, the krui tree produces the *damar* resin, a translucent lemon chiffon. As one forest farmer explains, "Damar trees are productive for hundreds of years. It is not heavy labor, like construction wage work or oil palm." Kayapo poly-cropped forest-gardens in Bahia, Brazil, have a mix of bananas, jackfruit, cocoa, and pineapple under close to zero input management: a bit of sawdust and lime, as opposed to nearby monocrop cocoa plantations which rely on extensive fertilizer inputs. The former produce equivalent coca yields as the latter, while on a per-land-area basis the "overall productivity of the forest garden is higher," and at the same time, creeks began to ripple across the area during the period under study, showcasing how such forms of farming began to improve the water cycle.[64] In Western Java, diverse agroforestry systems producing cajaput oil sequester far more CO_2 than monocrops.[65] Forest-based home gardens can lock as much carbon into the surface- and subterranean carbon cycle as secondary growth forests, putting to rest concerns about the need to recreate or lop off land for nature in lieu of humans in order to achieve maximum CO_2 drawdown.[66] Similarly, Brazil's Bahian cacao forest-gardeners seal great quantities of carbon in their soils through perennial agroforestry.[67] In India, scientists estimate a hybrid Napier and mulberry silvopasturing system on three million hectares would supply 72 million additional tons of dry matter and protein, a third of the deficit of the Indian livestock industry, and absorb CO_2 equivalent to 1/10th of India's current annual emissions.[68]

Agroforestry more broadly allows for unclear but massive replacement of cereals with perennials, which fix more CO_2. Banana and coconut intercropping yields more calories than rice (a major source of methane emissions, although methane is not the problem it is made out to be) and commercial apple production produces 24 million calories per hectare, whereas corn produces 14-30 million.[69] Breadfruit, a tropical staple, produces at least three times as many calories per hectare as corn.[70]

Agroforestry would be harder to implement in the US or other "First World" nations, primarily because it is more labor-intensive. However, as a way of managing the land it is not new at all, and was long part of Indigenous husbandry practices. Some researchers calculate that through five strategies – silvopasturing, or integrating trees into pasturing systems, or planting forage under trees, alley cropping, which inserts trees into fields, windbreaks, riparian and upland buffers, next to streams, and forest farming – the continental US could absorb CO_2 to the tune of 530 teragrams per year, or around 11 percent of 2019 emissions. Such measures bring with them a host of social and ecological benefits beyond their excellence in pulling gases from the atmosphere: habitat restoration, fire protection, increased incomes and yields, erosion control, watershed protection, diversification of income.[71] Silvopasturing moderates temperatures for grazing animals, and they require less forage, while trees can increase the amount of forage available to animals – doubling it under walnuts, for example. They also produce more value in price terms.[72] Indeed, through coppicing, in which trees are cut to the stump and shoots allowed to sprout out from it, the leaves of trees can be part of animal forage.[73]

Trees also can produce a lot of food in the North. Chestnuts produce two-thirds of the calories as wheat per hectare, and seed far more CO_2 in their trunks and roots and soil. Walnuts produce more oil although a little less protein than soy, with far more CO_2 storage.[74] In Illinois, modeling output based on Chinese chestnut, European hazelnut, blackcurrant, and a hay alley crop could reach more than half of the yield of modern maize/soybean plots when they are mature. Nor is this merely scratching and lab work.[75] The Savanna Institute is (re)introducing agroforestry in the Midwest with excellent economic, ecological, and social results. Yields on nut trees are based on far less agronomic research than has been devoted to industrialized cereal monocrops, which means a people-centered agroforestry research program would almost certainly lead to more nut production.

Such changes in farming imply partial shifts away from so much cereal production. This should not be alarming, for two reasons. One, certain types of trees or poly-cropping parcels produce as much or more per acre as cereals. Two, far more food is currently produced than human beings need, even while many farmers produce far less than they could if they were supported through low-tech extension and participatory research and breeding.[76] Cereal farming is so widespread because cereals produce a lot of caloric output relative to the amount of labor which goes into growing it, because it is pretty dense calorically speaking – although far from the densest – and because it stores well. And because under modern settler-capitalism, there are huge subsidies for it, and huge profits to be made from selling cereal farmers their fertilizers and seed, and buying their grains and corn and processing it into not very nutritious but highly calorically dense food.

A shift from annual cereal monocultures to more agroforestry in large parts of the world will be a large part of a transition to worldwide agroecological production. The substantial remainder of cereal farming should in no way be discarded, nor cereal farming necessarily minimized, but instead transformed. A wide range of experiments show that adopting agroecological methods will reduce northern yields of corn, wheat, and rice by about 25 percent, while, of course, cutting CO_{2e} emissions related to dolloping fertilizers, pesticides, and herbicides onto fields to zero.[77] Furthermore, taking a very different approach, a wheat relative has been successfully perennialized, as part of the Land Institute's 50-year plan to reconvert US and global cereal farms into consumable equivalents of the perennial prairies of old, with root structures meters deep, another way of doing agroecology which involves direct biological mimicking of prairie production with its huge elongated tangle of roots, the jet dirt underlying them, and the cacophony of biodiversity which sat atop them.[78]

AGRICULTURE AND MEAT

Although cereals are the major issue when it comes to what people eat, they are not the major issue when it comes to emissions. Here, animal agriculture is front and center, with rising demands for global veganism and the slicing out of animal flesh from the human diet. These requests come from liberals like George Monbiot, who insist that "Meat and dairy are an extravagance we can no longer afford," or self-styled Marxists who call for a half-earth strategy based on a "meatless society."[79] Whether these are calls for compulsory veganism is up to the reader. Claiming a meatless society is a good thing creates an immediate moral justification for doing things to stop people – especially pastoralists in the Third World – from ceasing to raise or eat meat. Perhaps more importantly, this is the agenda of many in the ruling class, as we saw in Chapter 1, a fact which goes almost universally unacknowledged amongst contemporary "red" vegan scribblers, but which speaks directly to the role such writers play in confusing and disorganizing leftist opposition to the explicit agenda of the ruling class.

Let us consider the arguments: is this in any way, shape, or form a good idea?[80] Start with the simple points of agreement. The conversion of grasslands and forests to croplands and the stuffing of cows full of petroleum-laced corn must stop. Confined animal feeding operations, where tens of thousands of pigs and poultry are packed together, should also be entirely eliminated.

But there are also disagreements: calls for command-and-control war communism-style ceasing of meat-eating are a horrific idea. First, they are ecologically illiterate. In the beautiful days of the bison, herbivores stampeded all over the landscape. They and other grazing animals are part of the ecology: "soil microbial and associated plant communities in grazing ecosystems did not evolve under abandonment, they coevolved as complex, dynamic ecosystems comprising grasses and soil biota, the grazers, and their predators."[81] Even if human-bred ungulates were removed, other large animals would probably fill in their niche, and we would have a very similar methane cycle. Indeed, because herbivores have long roamed over grasslands in the US and worldwide, it is odd that their burping and flatulence and manure should suddenly be the sign of human-made carnivorous devilry. The great meat purge almost certainly starts from the wrong baseline, asserting that the "natural" state is a dreamland ecology without herbivores, when in fact those estimating baseline methane emissions should try to figure out the historical level of methane emissions from herbivores and termites and start from there.[82] Before the colonial invasion, the level of methane emissions from bison, elk, and deer was around 86 percent of present-day emissions from "farmed ruminants" in the United States.[83] Furthermore, new methods of counting gasses are putting the anti-meat crusade on uncertain footing. Methane, unlike carbon dioxide, is a very short-lived gas. Even small reductions in annual animal-sourced methane based on small year-on-year herd shrinkage would very shortly lead to reductions in methane's overall global warming effect.[84]

Additionally, the seemingly neat category "meat" as a cause of emissions obscures the ecological, political, and social differences between Sahelian herders, Kansan artisanal animal husbandrymen practicing intensive rotational grazing, and those sitting atop pyramidal monopoly-capitalist cow plantations or the pig concentration camps of North Carolina. Each of these are fundamentally different forms of life.[85] The demand that meat-eating cease, coming from powerful academics and even more powerful institutions surveyed in Chapter 1, is a suggestion that people who are currently producing and eating meat stop eating it. Such a suggestion's effect, whatever its intent or its purported purely normative content, creates northern consensus around encouraging, coercively or otherwise, transformations of how people live. It also creates a justification for unnatural "climate solutions" based around biofuels and bioenergy or "afforestation" based on trash-tree plantations which will allow the great petroleum corporations to keep burning their assets, to great profit. And it will be the poor who will suffer. In South and Southeast Asia, most ruminants are the property of smallholders, who use them as draught beasts and for milk, meat, and to store capital. In semi-arid West Africa, livestock are crucial for the well-being of rural people, and in the Maghreb, animals are disproportionately the property of the poor and used in mixed farms.[86] The injunction to stop eating meat can also imply either the displacement or erasure of Inuit and Sámi people. There are, of course, few or no jobs for those people who would be displaced from pasturing, which would lead to further wage depression and further human suffering. These things can easily occur whatever the subjective intentions of advocates for "mandatory global veganism," all the more reason to focus on industrialized animal agriculture far more than that shapeshifting thing, "meat."[87]

Furthermore, some of the anti-meat mission rests on Malthusian notions that we are running out of resources, whether water, land, or feed. But as ever, Malthusianism rests on sloppy empirics. Dried apricots and almonds are the primary water-guzzlers in the Australian diet, for example.[88] And 86 percent of what meat animals eat is inedible by humans, and even in the US, most cattle eat grass for most of their lives. The idea of a great trade-off between animal food and human food relies on myths of insufficiency on the one hand and simple lack of knowledge of what animals eat on the other. This is another reason not to rely on the broken crutch of veganism to deal with the phantom injury of running out of food.[89]

Instead, animals can and should be part of landscape management, helping store CO_2 in soil, and enriching it and making it more resilient in the face of climactic perturbations, which will come with more climate change. There are agroecological options for the northern animal husbandry arena, aiming at integrated animal health management, and eliminating CAFOs or other hellish incubators for plagues rippling out from capitalist agriculture. We should increase species diversity within livestock systems as to increase resilience – remember, nature abhors a monoculture – preserving biological diversity, and building in firewalls against virus transmission that happen when all animals share the same DNA. Animals can and should be part of farms.[90] These best practices include integrated agro-pasturing operations, integrating poultry into farms, and a grazing technique that seeks to mimic how animals used to eat offer the most optimistic projections. These techniques are called mob or adaptive multi-paddock (AMP) systems in which animals eat the grass voraciously for a few hours or days, trample seeds and manure into the soil, then move on to the next paddock. These systems mimic the natural grazing of wild ruminants. They are also in use by the Maasai and Barbaig communities in Kenya and Tanzania.[91] They require active herd management rather than irrational colonial fences, and they allow grass to regrow before it is grazed again, in effect regrowing its solar panels, or its leaves, like a self-replicating organic machine.[92] Such systems produce dense and fibrous roots that help store CO_2 in soil. Such systems invert the balance of CO_{2e} emissions on a lifecycle basis, potentially turning animal tending from a net-positive to a net-neutral or net-negative sector in overall CO_{2e} emissions.[93]

We have many examples of animals being smoothly woven into complex production systems, with benefits for humans and the environment alike. In Spain, for example, the Iberian pig is the keystone piece of the *dehesa* system: human-made oak savannahs peppering the peninsula, rich with cork and holm oak, and producing cork, *cerdo iberico*, game, mushrooms, and honey. This agroecosystem is 4,500 years old, a revamping of human-made-with-fire nut tree savannahs from the Mesolithic period, and replaces wild boar and aurochs with pigs and cattle.[94] A study in southern Spain showed that on cow farms using a variety of sustainable methods, carbon sequestration was 89 percent of total CO_{2e} produced. For dairy goats, the total was 100 percent, and the same for Iberian *montanera* pig farms (these studies accept traditional methane accounting).[95] Experiments in Michigan have shown that for the "finishing" phase, adaptive multi-paddock grazing reduced emissions from 9.62 to −6.65 kg CO_2-e kg of the carcass weight, although these experiments also showed that this type of finishing required more land per unit of meat than other kinds of finishing. As the researchers concluded, this "challenges existing conclusions that only feedlot-intensification reduces the overall beef GHG footprint through greater productivity."[96] Experiments in the southern Great Plains echo these findings.[97] Because lands good for grazing are not necessarily, or at all, suitable for crop growing, AMP methods are not in a zero-sum game with food crops for people.

They also produce other ecological benefits. In studies involving bison in the upper Great Plains – yes, bring back the bison! – AMP produced greater water infiltration, improved land composition and forage availability, and a decrease in invasive plants. Water infiltration is particularly crucial in warding off storm-borne erosion due to the enormous and sudden rainfalls slated to increase in a warming world.[98] AMP also allows for higher stocking rates amidst climactic uncertainty, or more animals per unit of land.[99] Furthermore, carbon stored in rangelands is locked up far more securely than carbon stored in burnable tree trunks.[100]

What about the economics? In one French farm which converted from a high-input to a low-input system, nitrogen loads plummeted, pesticide applications fell by two-thirds, and energy consumption fell by one third. Milk production decreased around 5 percent but because input costs were far lower, the farms became more profitable. Other surveys show uniformly increased profits for farmers since they rely less on off-farm inputs, and even less labor needs.[101] In El Hatico reserve in Colombia, which has used these techniques for decades, stocking rates have increased and milk production has more than doubled, while chemical fertilizers are no longer used.[102] We know less about whether people in the US and Europe will have to reduce their meat consumption. It absolutely feels intuitively correct to claim that there will have to be reductions, possibly sharp reductions, in meat consumption in the core. Some claim that if meat were produced sustainably, people would need to eat up to 70 percent of it.[103] A few agronomists and herders estimate that the US could produce more than it currently does.[104] That remains to be seen. As with so much of what is or is not possible with farming which rides rather than wrestles with ecological cycles, we just do not know, in large part because we have never tried to have a farming system, or used techniques, or heavily researched techniques, whether in the US or the world, which aim to serve all people rather than the few and the powerful.

AGROECOLOGY AND YIELDS

Now consider yields and production, to establish that this farming mega-shift would not incite starvation. First, currently the world produces, even by conservative estimates, around 3,100 calories per person of available food. About half is grown using relatively CO_2 light methods, by smallholder farmers in the South. The other half is grown using CO_2 intensive technologies – pesticides, herbicides, soil tilling, no cover crops, and artificial fertilizers.[105]

Although the emphasis is on a northern People's GND, it's important to briefly note that food sovereignty, based on agroecology and radical land-to-the-tiller agrarian reforms, is the cornerstone for Third World eco-socialism. From an environmental perspective, agroecology pulls CO_2 from the atmosphere, without a need to invent a high-energy machine for doing so. Existing knowledge will do it for us, with estimates from La Via Campesina that worldwide agrarian reform coupled with agroecology could suffuse the soil in organic matter and vacuum CO_2 from the air in massive amounts.[106] Calorie yields from more labor-intensive polycultures on a per-hectare basis are not that much lower than monocultures. Monocultures do better when one compares, say, the yield of a hectare of corn doused in the most extreme growth-promoting fertilizers with corn which grows perpetually in a Mexican *milpa*, nourished only by nitrogen-fixing black beans – two of the "three sisters," alongside squash, which compose a balanced, healthy, non-carcinogenic and permanently sustainable diet for huge swathes of the Mexican *campesino* world. And the yield differences for crops like corn or wheat, the mainstays of industrial farming, are high but not too high, around 25 percent reductions.[107]

In Cuba, the beacon of agroecology, yields have continuously increased in major crops like beans, rice, plantains, and potatoes since 2007.[108] Agroecological techniques ranging from vermiculture, soil conservation, innovation of intercropping designs, seed saving and recovery of local races and varieties, peasant seed selection and crosses, participatory plant breeding, increased crop/livestock integration, improving of local animal feeds and pastures, have coincided with vast contributions from the peasant sector to national feeding. And furthermore, in Sancti Spiritus province, intensely agroecological farms as opposed to farms just lightly applying agroecology produced four times more per hectare and three times more per worker-hour.[109] And all of this occurs while rapidly reducing if not erasing CO_2 emissions from the food system and increasing sequestration.

Elsewhere, Mexican *chinampas* in the mid-1950s produced 3.6–6.3 tons per hectare using traditional farming techniques. If a hectare produces enough food for 15–20 people, one farmer working year-round could feed 12–15 people. Maize, squash, and beans – the "three sisters" of traditional Mexican *campesino* farming – can produce as much food on one hectare as 1.73 ha planted with maize under equivalent conditions. Peruvian bench terraces have increased potato and barley yields by 40 percent. In Bolivia, intercropping with Lupine produces potato yields of 11.4 metric tons per hectare, at far higher energetic efficiency and improved income for the farmers as even though yields are lower than in "modern" high-input systems, costs are far lower, while the "externalities" of decreased soil quality and use of non-renewable fertilizers evaporate. In Brazil in Santa Catarina, small farmers have used green manures, contour ploughing, and intercropping and have increased maize yields from 3 to 5 tons per ha, and soybeans from 2.8 to 4.7 tons per ha, while soils are darker and brimming with biological activity, and reduced weeding and ploughing needs have meant increased yields go along with lower labor inputs.[110] Farmers in Bangladesh have seen 4–11 percent yield increases in rice, alongside sharply reduced yield oscillations and 25 percent reductions in cost, while producing fish in the rice fields and vegetables grown on dikes has increased farmer income.

Fish and vegetable consumption has doubled for many families.[111] In El Paraiso, Guatemala, drainage ditches and chicken manure applications increased bean and maize yields by 700 percent, sometimes over five tons/ha for the latter, and allowing families to shift to fruit, vegetables, and coffee – and in some cases dramatically out-producing kindred farms which relied on chemical fertilizers. Soil became rife with earthworms, with overall biodiversity blooming.[112] At least, until chicken farmers charged too much for the farmers to be able to afford the manure.

In Andhra Pradesh, India, farmers using the Zero Budget Natural Farming (ZBNF) method of mulching, *Bijamrita*, a treatment of cow dung and urine to seeds and seedlings, and *Jivamrita*, liquid and solid inoculants to head off fungus and increase soil organic matter, have increased yields by around 16.5 percent and decreased production costs for rice, maize, groundnut, and millet, leading to 50 percent increases in farmer incomes alongside increased ability to resist the shocks of droughts and cyclones. Increases in living creatures in the soil have been astonishing: in comparing ZBNF yields with non-ZBNF yields, researchers found 232 versus 32 earthworms per square meter, as well as huge numbers of beneficial insects, from pollinators to pest antagonists, from honeybees to lacewing bugs, "an antagonist to aphids, leafhoppers, whiteflies and mealybugs," as well as ladybugs.[113] In Tripura, India, the system of rice intensification using agroecological techniques for fertilizer, water management, and weed management, doubled yields and had vastly increased returns-to-labor-and-cost, all while boosting the ecology.[114] Experiments with manure applied during double rice cropping showed this is the best way to increase carbon matter in the soil, which means leaching it from the atmosphere.[115] In Karnataka, Orissa and Madhya Pradesh, SRI increased yields by 99 percent and incomes by 109 percent.[116] In Kenya, Uganda, and Tanzania, the "push-pull" pest control technique, which uses plants that repel pests from crop fields and attract them elsewhere, increased maize yields from 1 ton/ha to 3.5 tons/ha and sorghum from 1 to 2 tons/ha.[117]

These are southern miracles, but they relate to a People's GND in the North. They make clear that the South is capable of more than feeding its own populations, if there are radical agrarian reforms and state support for agroecological research as well as strong state infrastructure to support these efforts. Indeed, it seems likely that a people's ecological-agricultural socialism in the Third World would produce more food, higher incomes, on the same amount of land, and using less labor. US/European food surpluses are not needed. In fact, they have been harmful to Third World food systems.[118]

A CARBON-FREE US FOOD SYSTEM: SOME EXAMPLES

What kinds of shifts would a carbon-dioxide-negative agroecological US food system imply? The statistics I cited above about shifts to agroforestry and conscientious grazing show that an agriculture which does not go alongside, but *is* land management, is possible. Modeling from Britain done by Chris Smaje projects that even with abysmally low cereal yields, 25 percent of current ones, Britain could feed a population of 20 million more people than it currently has, using just 15 percent of its labor force, or seven million agricultural workers out of a projected 83 million people in 2050.[119] Roughly mapping those figures onto the US, that would be around 23 million people. That sounds like a lot – way too many, I think, absent the collapse of modern social systems, which is beyond my scope here. But that's OK: a People's Green New Deal should in no way advocate putting unwilling people into the countryside. Instead, consider there are around 2.6 million farmers in the US, plus 2-3 million migrant agricultural laborers.[120] Consider furthermore the time US citizens currently spend on backdoor or community gardening or lawn care, usually on their own dime: around ten minutes a day, per adult. Imagine that was self-directed labor for which people received a social wage, and it is easy to imagine scaling up that number. These numbers do not remotely mean everyone would be involved in food production. They simply demonstrate that there is a social basis upon which to make land-healing labor, including environmentally restorative agroecology, fully as attractive if not more attractive than other kinds of work. Such a decision would need to be established through comprehensively planning how much labor is needed to carry out land restoration and agroecological farming, making sure such work was fairly distributed, and working from there.

Of course, the dream of a master-planning state is quite far from the current capacities of anti-systemic forces in the North. Agrarian municipalism is a more likely possibility, made more feasible once we note that food ideally anyway ought to come from local foodsheds, mapping roughly over the watersheds, which are the ecological basis for a more bioregional vision.[121] Indeed, soil itself is profoundly local:

> By creating locally or regionally unique crops with their own "terroir", evolutionary breeding is in line with a re-connection between producer and consumer on a regional level...For this link to be made, a requirement is that the evolutionary breeding process should be de-centralised, i.e., that populations are developed on-farm rather than being maintained centrally and distributed to multiple locations.[122]

Examples abound of agroecological experiments in the imperial heartlands: The Savannah Institute patiently developing agroforestry systems and the Land Institute in Kansas experimenting with grain perennials. The Sustainable Iowa Land Trust trying to decommodify land. Cooperation Jackson is another one, part of a very long heritage of Black nationalist agrarianism, which has forever been woven into the fabric of Black revolutionary activity.[123] A third is Soul Fire Farms, using Haitian and Namibian farming techniques to feed the people, build up the movement, and train farmers. A fourth is Hawai'i's 'Ike 'Ai Consortium, working to build up a solid and secure foundation of locally owned and family-sized farms which used ecological production practices and local inputs, paired with the "biocultural restoration of traditional Hawaii land and seascapes," including its low-tech masterworks of stone fishponds.[124] In the current context, more striking than the small numbers of people involved in food production is how many groups and organizations and collectives are devoting their lives to a form of labor dismissed as patriarchal, laborious, back-breaking, or otherwise despicable. Consider the words of Leah Penniman of Soul Fire Farms, who states that after a summer working at a farm, "I had made something and had fed some people with my hands. I think there's so much we're trying to do, but fundamentally it comes back to the fact that all of us want some dignity and all of us want to be able to produce something of value and to be something."[125]

WHAT IS TO BE DONE?

The foregoing suggests the problem of an agroecological transition is not yields, nor is it ecological rationality. It is probably not the absolute impossibility of enough people finding the work "objectively" likeable. Rather, the obstacle is structural: it is capitalist and imperialist organization of the food and agriculture system. The solution is anti-capitalist and anti-imperialist structural change.

The most basic agricultural elements of a People's Green New Deal are:

1) The dismantling or nationalization and retooling of the large corporations involved in agriculture, like ArcherDanielsMidland and Monsanto. They should become worker-owned laboratories for agroecological innovation and appropriate-scale mechanization.
2) Large-scale agrarian reform, breaking huge farms into units which can be tended by families using agroecological methods, or lassoed into cooperatives.
3) Parity pricing, so that farmers do not need to overproduce, and so that prices give a good return to labor, so that people have excellent lives in the countryside, and so that the ecological benefits of non-industrial farming are fully reflected in the pricing system.
4) Abolition of the subsidies which support industrial agriculture, including those which support monocultures, and even more those monocultures used for biofuels and animal feed.
5) Massive investment in a green transition to get farmers "over the hump" of a transition to agroecological production.
6) Public investment in all the needed infrastructure to re-localize food systems, including local processing, abattoirs, and other needed physical plant. Such a move would also empower local non-farming labor.

In a People's Green New Deal, high-quality social housing, healthcare, and organic food are decommodified and are rights. Social planning to compensate people for adding to social wealth through creating use values or restorative work on the "free gifts" of nature is feasible. And would people wish to do it? Consider what people do with their free time, when they are not coerced: community and home gardens when possible. Artisanry and art. Such programs could be part of an eco-socialist transition alongside increasing elimination of private ownership of the means of production that make food and housing and healthcare only accessible through the price nexus. The People's GND is also a call for seed sovereignty. Genetic material should remain in the hands of the people and not in the hands of corporations. And of course current US and global maldistribution of land is intolerable. The US and the world more broadly should have their own agrarian reforms, based first of all on land back, or decolonization of the settler land base.

Go on to soak the countryside in culture. At least one aspect of the information revolution has been beneficial: democratic and decentralized access to culture. In a flash, the entire encyclopedia of human knowledge is available to everyone on Earth, at least in principle. Promote massive and democratic access to free high-speed internet on a global basis. Secure full and equal access to electricity. Secure and respect sovereignty for Third World countries, which will redistribute land, cease using it for export crops, and be able to lock in surplus locally, including through locally sovereign agroecological research lighthouses, now beaming hope from southern India's Zero Budget Natural Farming, to Havana, to La Sociedad Científica Latinoamericana de Agroecología in Colombia, to the Agro-Ecology and Research Corps in the US. Technology transfer should include grants for further development of national-sovereign agricultural research and extension systems (and this is where we return to the national question, since land-to-the-tiller agrarian reform has been the surest way to attract the attention of US capital and its military). Such a shift must also include breaking the "free trade" steals which shatter southern agricultures and allow for tariff walls and subsidies based on nurturing food sovereignty, not chasing the chimera of food security.

For the South, such an agroecological revolution would be a gift beyond dreams. But not just there. In the US's domestic "Fourth World," there is a bloom of interest in "resilience gardens"[126] amongst native communities, to youth growing heirloom guayaba, maiz, and frijol in South Central Community Garden, where:

> [f]or thirteen years, the community – including native peoples of Mixtec, Tojolobal, Triqui, Tzeltal, Yaqui, and descent – has relied on a rare piece of urban open space to grow food while becoming self-reliant and building a sense of community. South Central Farmers Feeding Families is a grassroots organization of 360 families.[127]

Many of those communities, to be sure, are no strangers to struggling against apocalypse. Nor, on the other hand, are these struggles purely or even primarily defensive. They are also, potentially, the dress rehearsal for a better world, a wattle of institutions which could be the basis for an entirely new way of organizing not only the US food system, but the US as a society – a way of building up a food system which is not vulnerable to virus and blights, which safeguards the health of the poor, and which does not demand the exploitation of southern lands and peoples. Indeed, writes farm leader Jim Goodman, "We have always had the solution." The challenge is taking it up.[128]

7

Green Anti-Imperialism and the National Question

The foundation of a People's Green New Deal is the national question: the right of self-determination towards political and economic sovereignty. The national question is the set of political problems concerning oppressed nationalities within nations, colonialism, self-determination, and national liberation. The "question" surfaced historically throughout the history of the socialist movement. In particular, it arose as the 3rd International, led by Lenin, was trying to achieve a principled policy towards the colonized and dependent nations. In this context, the national question was a way of understanding the political topography of imperialism. One central "national question" emerges out of the "distinction between the oppressed, dependent and subject nations and the oppressing, exploiting and sovereign nations." The second set of nations carried out the "colonial and financial enslavement of the vast majority of the world's population."[1] That is, the national question articulated with the class struggle and the search for popular emancipation. Today, the national question takes a variety of forms: ongoing struggles for decolonization and self-determination towards political and economic sovereignty within settler-colonies like Israel and the United States.[2] And furthermore, struggles to stanch South–North value drain are also part of the national question, as are struggles which connect to combating the flows of value which are the essence of the relationship of oppressing and oppressed nations.[3]

For those reasons, the national question arises in South and North alike. It is the foundation of a People's GND. This might seem an odd foundation stone for a people's agenda in a settler-colonial empire. The United States rides roughshod over the sovereignty of other states. But that is precisely the point. Other peoples' national questions, especially those of Indigenous peoples, must form the basis of a People's GND within the territory of the United States and in other settler-colonies, since building ecological societies requires popular control by the most excluded over their national productive resources.

It can be all too easy to sidestep or suppress the national question, in favor of a sort of easy ecologism, or a compression or reduction of fundamentally political questions to environmental management. To that end, one response has been to argue that reducing the rate at which the imperial core spills out CO_2 and other environmental toxins into the world is internationalist by definition, and that we had therefore better focus on that mission. Many then suggest that degrowth in the wealthier world, which would reduce its material impact on the remainder of the planet, is the most effective internationalism, leaving more space for others to live.[4] But there is a thin line between modesty and myopia, an inwards-looking ostrich syndrome, in a country marked by imperial modes of living.[5] Such a move may reduce political choices and demands about paths forward to simply retooling the machinery of growth. Such an internationalism can inadvertently silence demands for climate reparations. A similar but more Eurocentric arrogance claims that an ecological politics for "the working class" means a politics for the northern industrial and service sectors (cotton farmers in India, pomegranate farmers in Iran, phosphate miners in Tunisia, and many others do not in this view figure as "working class" but as odd curios who somehow are not part of the world capitalist system and its prices, even though the capitalist system pays plenty of attention to some of them via sanctions). This approach disdains any transformative aspirations. It effectively erases ecological debt, offering another version of green pseudo-social democracy in its praise for the Markey/Ocasio-Cortez GND.[6] A third approach looks to supply chain justice and cross-national worker solidarity for just transitions "from below" as the basis for developmental convergence and unity-in-difference, while equally eliding entirely the historic language of climate debt, and transmogrifying the national question into aesthetic gestures about imperialism and colonialism.[7] And a fourth and final approach, coming from institutions like the UN Conference on Trade and Development,[8] imagines "ideologically foist[ing] Northern ideas and framings of 'green transitions' centered on techno-optimist imaginings about renewable energy futures which are either unsuited to the realities of the global south, do not center justice, [or] have any serious Global South consensus."[9]

While supporters of degrowth are the most sympathetic if not outright supportive of payment of climate debt, and are the most open to Third

World Marxism, other internationalisms sidestep serious engagement with the past, present, and future of a hierarchical international system organized around nation-states. These other internationalisms deny that polarization is inherent in the capitalist world-system, green or otherwise. And they reject the national question as necessary for organizing thinking about autonomous forms of resistance towards emancipatory horizons. Discussion about GNDs in the United States are never – and can never simply be – about the United States, because US capitalism is not and never has been just about the United States. US wealth was built on a continental process of primitive accumulation of the land, and continuous wars which transmute lost lives into the stock market valuations of US corporations.[10] Its circuits of accumulation cut across borders and crosshatch the world. Its long tradition of burning cheap petroleum and coal to build up an immense and convenient infrastructure came at the cost of denying other countries the capacity to opt for the same resource-use path. That history of large-scale burning of resources cascades into developmental disarray and de-development even now, from millennial cyclones in Mozambique to deluges in Bangladesh and the looming submersion of the Seychelles. And future decisions about how much lithium to use, to pay or not pay climate debt, and how much energy to allot to each inhabitant of this territory inevitably reverberate globally.

THE NATIONAL QUESTION: THEORY

In an imperialist world, environmental politics have a specifically national aspect.[11] Because imperialism, the transfer of value from South to North and the uneven development accompanying it, continued long after the end of the major wave of formal decolonization, the national question is not a historical relic, antiquated and anachronistic.

First, colonialism itself is not over. It endures de jure in a plethora of settler-states, and its afterlives haunt the periphery.[12] As formal and legal decolonization gave way to neocolonialism, nations lost control over their economic sovereignty, the pot of gold they had hoped to find at the end of national liberation. Nation-states, the political containers through which accumulation on a world scale and uneven development endure, persevere, and deepen, structure uneven access to the fruits of world production. Countries like the Democratic Republic of Congo, Iraq, Venezuela, and Yemen face national losses in their productive forces, through sanctions and the threat of war or war itself. For these reasons, the national question endures. On the other hand, the nation is one of the political and social units within which people organize to resist oppression. A large number of the most dynamic struggles from the 1980s to the present, from Venezuela to Bolivia, have used a national-popular idiom to articulate their politics and place domestic wealth at the service of the peasantry and working and marginalized classes.[13] It is perhaps underappreciated that the most globally cherished struggle for liberation and justice in today's world, that of the Palestinians, is that of a subjugated *nation* fighting for land, liberation, and return. And there is no hope that Palestinians – or Yemenis – will receive and control climate debt reparations unless they have de facto and de jure national sovereignty, the political shells within which thinking about the future can occur.

It is in and through the national political sphere that decisions about rates of investment and disbursement of social goods must be made, alliances built, and internationalisms constructed. It matters, for instance, who is at the helm of the Bolivian state. It was Bolivia, a sovereign and national-popular Indigenous-led nation-state, that was the sanctuary for the Cochabamba documents, which emerged from meetings in late April 2010, and demanded wide-ranging payments from North to South alongside a radical climate agenda to meet the needs of Mother Earth and poor humanity alike.[14] And it is through the state system that ecological debts are calculated and demands for debt repayments are made in world political fora. Focusing on the national question underscores the right to regain control of a people's historical process, for people to decide how and with whom they want to live, and not to have that decision made for them by a superior class or a more powerful colonizing state. And this includes Indigenous peoples, which cannot be reduced to any kind of beneficiary of a restored prelapsarian ecology, but who are, as Indigenous scholars Andrew Curley and Majerle Lister write, emphatically "modern peoples whose greatest threats are political marginalization at the hands of continued colonial processes."[15]

Insisting on the importance of the national question does not require denial of social or democratic questions: who gets what within nations, who decides who gets what within nations, and who gets to shape the environmental architecture of production and distribution within and across nations. Such an insistence simply reflects the hierarchical structure of the world-system. The United States' assertion and exercise of extra-territorial sovereignty has created a significant "sovereignty deficit" in many other states, limiting their power and authority. This is a continuous feature of settler colonialism, pacted decolonization – or decolonization which occurred only through extensive dialogue with the colonial force and led to significant surrenders of national sovereignty – and neo-colonialism, which has reduced the physical resources available for poor people to build up their own lives.[16] Indeed, even during the brilliant noon of decolonization from 1947 to 1980, farmlands, forests, banks, currencies, factories, salt, iron mines, quarries, and oil fields remained in the hands of the colonizers. Almost never was decolonization so successful as to allow peoples to fully determine their own histories, even within their own nation-states.

Furthermore, the national question is a coin with two faces, imposing distinct political and social burdens of transformation, planning, and struggle in the North and the South, including the "Fourth World" of Indigenous peoples.[17] That is because rights are neither possessions nor abstractions. Rights are relationships. Any right of the Third or Fourth World implies First World respect for that right, including the political struggle to secure such respect. That is, rights imply responsibilities – Sioux rights at Standing Rock meant people of all kinds had to go and stand with the Sioux, Lakota, and other Indigenous to fight for those rights. If one believes that Palestinians have a right to national liberation and self-determination, there is an implied obligation for everyone, from their own specific locations, to assist that struggle, including identifying the manacles forged by core nations that enchain the Palestinians. Whether GNDs take the form of local planning or a new global architecture for just transition, all parties must shoulder the burden of that transformation, which involves assessing how much the First World as it currently exists – its skyscrapers, mass transit systems, beautifully wired metropolises of marble and granite, and a countryside increasingly sanitized of ecologically ruinous industrial production – has been based on a relationship which denies many rights to Indigenous peoples and peoples on the periphery of the world economic order, and trying to make amends for those denials.

At least three elements of the national question are central to just transition. First, the push to have the concepts of climate and ecological debt taken seriously. Second, movements in favor of demilitarization and the construction of a peacetime economy in the metropolitan core. Third, struggles against settler colonization, which are connected to attempts to reinvigorate sovereignty and safeguard our common home: the world and the global environment. These three elements overlap in their promise of an entirely different world. Demilitarization redirects social spending away from destructive and towards productive (and even creative) endeavors, liquidating the material foundations for the core's denial of peripheral sovereignty and popular development. "Land Back" projects do so too, since land is the principal physical basis for decolonization. By definition, reinvigorated sovereignty means Land Back. Ecological debt is the means whereby sovereign and suitable industrialization, and increasingly humane and popular development itself, becomes possible for peripheral countries looted through generations of colonial and neo-colonial drain. It is the organic fertilizer which allows economic sovereignty to bloom.

CLIMATE AND ECOLOGICAL DEBT

The concept of ecological debt is based on the diagnosis that capitalist production and consumption have vastly overrun the world's space for waste, including the atmospheric space for that all-important by-product of fossil capitalism, carbon dioxide. The concept of climate debt concerns the appropriation, or enclosure, of the world's capacity to absorb greenhouse gases, with staggering implications for the development prospects and pathways of the world's poor. Some also refer to what is often termed "adaptation debt," the resources needed for poor countries to control or otherwise respond to rising sea levels, increased typhoons, and other outgrowths of ecologically destructive capitalism.[18]

Settling climate debt is a material implementation of the international legal principle of "common but differentiated responsibilities," which stipulates that all states are responsible for addressing the destruction of the global environment, but they are not all equally responsible. Economic and other disparities between countries must be considered when designing legal obligations or responsibilities appropriate to their economic resources and institutional capabilities.

The frequent erasure or muffling of calls for climate debt from much northern climate discourse is inseparable from the contemporary rise of social democracy, with avatars like Jean-Luc Mélenchon, Jeremy Corbyn, and Bernie Sanders. Accompanying policy intellectuals have put forward a climate discourse which largely is silent on climate debt. And if social democracy in its classic form has been a prophylactic against revolution, it makes sense that climate debt, a Third World working-class demand par excellence, has been largely absent from most GND manifestoes. For while genuflections to internationalism abound, and programs refer frequently to settler colonialism, global racism, and apartheid, beneath them lies a denial of responsibility for colonialism, neo-colonialism, and imperialism.

These new positions are in stark contrast to previous left climate politics. Over a decade ago, the Cochabamba people's process took place, setting the framework for climate debt discussions. That meeting devoted a working group to tackling the issue of climate debt, building upon two of the core principles of the United Nations Framework Convention on Climate Change: the principle of common and differentiated responsibilities, and the principle of equity. This multilateral treaty, first opened for signature in June 1992, reflected state-of-the-art political and scientific understandings of a remarkable array of different topics: climate change; uneven incorporation into the capitalist world-system; the complex and multiple legacies of colonialism; the state-based organization of global politics; the sovereign rights of states to develop their resources; and these states' obligations to ensure that such rights were not exercised in such a way as to harm their neighbors "beyond the limits of national jurisdiction."[19] The 1992 Framework Convention declared that the climate must be safeguarded:

> [f]or the benefit of present and future generations of humankind, on the basis of equity and in accordance with their common but differentiated responsibilities and respective capabilities. Accordingly, the developed country Parties should take the lead in combating climate change and the adverse effects thereof.[20]

Countries listed in Annex I of the 1992 treaty were industrialized members of the Organization for Economic Co-operation and Development (OECD) or "economies in transition" (i.e. undergoing transition from state socialism). These countries were categorized as having historic responsibility for emissions and were required to commit to specific emissions reduction targets. Bolivia and other so-called "petrostates" committed to "extractivist" industries and seeking to lift millions of people out of poverty are not categorized as Annex I countries. Nor is China.

Building upon the treaty, the Cochabamba working group laid out a five-point program based on honoring climate debts, focused not merely on finance, but on "restorative justice," or "[a] means by which all peoples – particularly those who are mainly responsible for causing climate change and with the capacity to correct it – can honor their historical and current responsibilities, as part of a common effort to address a common cause. Ultimately, the compensation of climate debt is about keeping all of us safe."[21] An internationalist and eco-socialist updating of "from each according to their ability, to each according to their means," and an extremely material and working-class demand: what do oppressed people want and need more than to be safe in life and home? The five key demands were, first, the Olympian task of returning "occupied" atmospheric space – that is, to "decolonize" the atmosphere by reducing and removing emissions, to allot atmospheric space fairly, and to account for dual and potentially dueling needs for "development space and equilibrium with Mother Earth." The second such demand was to honor the debts that reflect lost development opportunities, since the cheap development paths blazed by the wealthy countries to build up their infrastructures cannot be walked again by poor countries. The third was to honor debts related to the destruction caused by climate change, including lifting migration restrictions. The fourth was to honor "adaptation debts," the costs of providing people with the resources to stay at home and have decent lives within their own countries. The fifth and final demand was to repudiate all efforts to segregate the climate crisis from the broader ecological crisis, and to honor adaptation and climate debts as a promissory note on the "broader ecological debt to Mother Earth."[22]

The history of battles over international debt at climate change-related conferences is the history of battles over the poor's right to the future. In Bolivia, the radical leadership of Evo Morales and Álvaro Garcia Linera was deposed by a US coup. (Their party has since returned to power in a brilliant display of popular organization.) That coup was preceded and made possible by a discourse of their ecological mismanagement, a mixture of outright falsehoods and quarter truths about the burning Amazon. Indeed, on a world-scale, Bolivia's environmental programs are not damaging but dazzling. Part of Bolivia's contributions to the fight for humanity's future was a call for far larger transfers of technology and financial resources than have traditionally been considered. Their proposals were designed to curb emissions in developing countries while realizing their right to development.[23]

A Swedish social scientist, Rickard Warlenius, has taken the Cochabamba positions as the basis for quantifying the climate debt. He argues that if space in the atmosphere had been calculated fairly, on the basis of how much CO_2 could have been safely emitted and absorbed by sinks, "the

North" (or Annex I countries) would only have emitted 15 percent of what it actually has emitted. "The South," by contrast, could have emitted more carbon dioxide than it has so far emitted, but not a lot: only about 4.4 percent more in total. In numerical terms, as of 2008, the North had over-emitted 746.5 $GtCO_2$ (gigatons of carbon dioxide).[24] As of 2008, at a carbon price of $50 per ton of CO_2, the value of the historical carbon debt would have been around $37.325 trillion. The International Panel on Climate Change estimates that a carbon price of between $150 and $600 is required to keep global warming below 1.5° Celsius. Those numbers would increase the size of the debt owed to the South enormously – to $111.975 trillion, at the low end, or to $447.9 trillion, at the high end.[25] More concretely, Bolivia has demanded "[p]rovision of financial resources by developed countries to developing countries amounting to at least 6% of the value of GNP of developed countries, for adaptation, technology transfer, capacity building and mitigation."[26] The GNP of the United States in 2019 was $21.584 trillion, 6 percent of which is $1.29 trillion. In that same year, the GNP of the entire OECD, comprising the bulk of Annex 1 countries, was about $54 trillion, 6 percent of which is $3.24 trillion per year.

Two facets of those mind-bogglingly huge numbers are especially important. First, they indicate that the former Third World taken as a collective, including the mammoth China, had clean hands c. 2008 when it comes to the climate crisis (although China's continued emissions since then change the picture). It had not emitted CO_2 beyond its fair share of the environment's capacity to absorb it. So, the climate crisis is basically the child of northern imperialism, pure and simple. Second, 6 percent of GNP per year far exceeds annual "growth" in the industrialized world. Although numerical indicators of growth are messy and map untidily over increases in the amount of physical material usage per unit of GNP, it is clear that considerable quantities of assets, technology, and rights (to timber, water, farmland, wheat, pomegranates, shawls, and transistors, the things for which GNP stands) would pass from North to South under such an arrangement.

The numbers in the Bolivian proposal are staggering – and probably deliberately so. They are compatible neither with capitalism nor with a polarized and highly unequal world-system. They are the arithmetic proof of the need for worldwide ecological and socialist revolution, and a revolution in North–South relations. Indeed, climate debt has been described as a "bomb."[27] The violence of the metaphor is appropriate because it is very hard to imagine a world system based on polarization between the South and the North enduring amidst massive debt payments going from North to South. And because these numbers cannot be argued with or dismissed, the North has generally ignored them, or tried to muffle and stifle them through the most effective way possible, a coup d'état. If the South has no strong and sovereign states, but merely neo-colonies, climate debt loses some of its most powerful champions and social agents.

DEMILITARIZATION AND A PEACETIME ECONOMY

For any serious response to the climate crisis, a key arena of structural transformation in the North is the removal of the US military from its role as a global police force and the United States' concomitant conversion into a peacetime economy. That shift must be part and parcel of the augmentation of southern sovereignty, the furtherance of decolonization, and widespread acceptance of the need to take climate debt seriously. Such struggles are interwoven, composing the social and political fabric of profoundly systemic changes.

Aside from the fact that the US military is a means of constraining the sovereign power of southern states, it is a tremendous fountain of CO_2 pollution. If the US military were an independent state, it would be the 47th largest polluter in the world, slotted between Peru and Portugal.[28] It produces uncounted quantities of waste the world over: heavy metals and pollution in waterways and grasslands. Jet fuel contaminates drinking water. Military bases are frequently Superfund sites – locations polluted so thoroughly that they require a long-term process of cleaning up hazardous materials. White phosphorus and depleted uranium are left behind by US weaponry in Fallujah, Raqqa, and the Gaza Strip, with a dark bloom of birth defects. The Grants Mineral Belt in New Mexico remains one of the best endowed uranium deposits anywhere, providing the raw material for the world-killing weapons. The mining takes place close to homes and communities of the Diné people, and those who have labored in the industry include Diné, Acoma, Laguna, Zuni, and Hispanic peoples, forcing upon them the social and ecological costs of US colonial-capitalist domination.[29]

It is urgent that the US military be eliminated as a major producer of waste in every sense, and to retool the labor it employs and the industrial plants it keeps occupied into socially useful goods, like solar panels and wind turbines and high-speed trains. This demand connects to the timescale of change. If far too much CO_2 has already been spilled into the atmosphere – and there has been, needless to say, far too much death at the hands of imperialism and capitalism – then the Pentagon system of global security must be eliminated immediately. The Pentagon produces not merely physical waste, but the waste of destroyed human lives. This churn of waste is the flipside of accumulation on a world scale. Furthermore, much of US industrial capacity should be devoted to producing clean technology for global systems change. As a fundamental demand of the climate justice movement, the Pentagon system and its sprawling manufacturing nodes ought to be converted to clean-tech plants.

In this context, it is worth recalling the appeals to security in the Markey/Ocasio-Cortez "eco-socialist" GND. Indeed, that GND is virtually silent on the national security state. Similarly, but also dissimilarly, consider Bernie Sanders, who recommended "[s]caling back military spending on maintaining global oil dependence."[30] In fact, Sanders called more broadly for redirecting the $1.5 trillion in annual global military spending towards a clean fossil fuel infrastructure. But that number was not the basis for policy.[31] Meanwhile, the incoming Biden administration wallows in national-security appeals: "regional instability…could make areas more vulnerable to terrorist activities," it warns. The response? "Invest in the climate resilience of our military bases," which will be needed in case there is a need for "military response."[32] The US Green Party GND has been clearer and firmer, advocating that the Pentagon budget be halved from its current titanic one trillion to a paltry $500 billion.[33] The Indigenous grouping The Red Nation is bolder still, declaring that "capitalism-colonialism on a global level" must be brought to an end, and that this includes "divestment away from police, prisons, military, and fossil fuels."[34] These differences matter. They have to do with whether the GND will be a vehicle for green social democracy in the United States, alongside superficial and largely tokenistic forms of anti-racism and "decolonization," or whether it will instead belong to a broader anti-imperialist program, involving the decolonization of settler-states in the Western hemisphere and elsewhere, alongside opposition to imperialism and the Pentagon system.

SOVEREIGNTY

Latin America, the Arab region, and Africa suffer under foreign intervention from the imperialist core, which plainly retains the right to determine who should rule other peoples, as it has done for centuries. Under colonialism, peoples were denied their right to history – their right to control their historical process and their social and economic life, including what to do with the land, mines, and trade of their territories.[35] The consequence of this denial of history in the nineteenth and early twentieth centuries was often late Victorian holocausts and massive wealth drain and colonial famine.[36] People were super-exploited: they were paid wages below those needed for their day-to-day survival and prospering.[37] The anti-colonial struggle, in the words of the outstanding anti-colonial theorist and leader Amilcar Cabral, was the "national liberation of a people," the "regaining of the historical personality of that people. It is their return to history through the destruction of the imperialist domination to which they were subjected."[38] And such domination did not end with the extirpation of formal colonial rule.

Even in the eyes of those who had led the most glittering and brilliant struggles for national liberation, formal decolonization was only an initial step. Shattering direct colonial shackles shifted the political apparatus. But it often left the economic apparatus under the de jure control of the "former" colonial power.[39] Across the South, many understood that independent development – in the Arab region, *al-tanmiya al-mustaqila* – was the logical continuation of breaking the colonial apparatus.[40] Struggles over the contours and shape of development stepped into the space previously occupied by calls for liberation.[41]

The world counter-revolution soon unfurled from the imperial core to the Third World and the heartlands of actually existing socialism. It dispelled dreams of development, most of which imagined, as did the Bandung projects, a non-aligned third way for the Third World, national-capitalist development within, against, and beyond the world-system.[42] Indeed, wars, invasions, hot capital flows, financialization, and coups burnt such hopes to ash. Or the heat of US global counter-insurgency warped even the steeliest nationalist-communist projects, setting some on the way to "market reform" and "market socialism," inciting full-bore Thermidor in others.[43] These national liberation struggles and non-aligned socialists were, for the most part, far too radical for the imperialists and not radical enough to ride out the imperialist storm (the lodestar of Cuba, the most ecologically advanced state on the planet, has been exceptional in its resilience).[44]

Slowly, imperialists, through the financial hook of neoliberal reforms and adjustment or the violent crook of war, sanctions, siege, and coup undid these governments.[45] They rejected their right to set their political and economic paths, to choose their alliances and friends, to weigh in on colonialism elsewhere on the globe, to carry out redistributive reforms, to take land from settler-thieves. As the 1990s dawned, rights to protect and intervene, responsibilities to protect, and the rest of the cant of imperialism had become the dominant chatter of a new "civilising mission."[46] In the dawn and dusk of the 2010s, increasingly, accusations of ecological mismanagement and extractivism soften up Western public opinion for the twenty-first-century's gunboat diplomacy and coup d'états.[47] By the dawning of the third decade of the second millennium, there exists very little substantive right to self-determination or sovereignty for the Third World amongst large swathes of Western public opinion. Of course, no one says as much. Such rights are formally accepted. But the notion that the US and EU must be prevented from abridging those rights through Western peoples taking on the burden of transformation by actively preventing their states from exercising unilateral coercion is scarcely present in Western politics.

A major element of the emerging imperialist consensus is the political organization of extraction from nature: green accumulation, environmental service commodification, monocrop tree plantations, and biofuels.[48] This occurs by gutting national states and reconfiguring their inner mechanics to turn them into conveyor belts for northern capitalist interests, narrowing "the scope for a green developmental state that could design a just transition to low-carbon economies."[49] When southern states have to pay off foreign investors when they exercise their sovereignty through nationalization or heavy taxes, this represents a shrinkage of sovereignty. It is turned into faded words on tattered paper, as its substance returns to the monopolies and the northern nation-states with which they are intertwined.

The struggle against such institutions and classes requires complementary and interlacing South–North struggles to open political space for and prevent reprisals against southern states which nationalize their productive plant, emplace humane wages, widen the scope of national control over productive forces, and ensure popular classes within the nation are the ones to exercise that control through their increased social power. Sovereignty must be strengthened because states are necessary vehicles for popular interests at our current historical moment. This is especially the case when it comes to climate, for which legal covenants take the state as central. Countries cannot demand let alone receive climate reparations unless they act freely and collectively on the world stage to get them. From the South, that means national-popular environmentalism of the poor and eco-socialism. From the North, it means acknowledging and struggling against state violation of southern self-determination. It is not meaningful to discuss debt payments without respect for the sovereignty of the countries owed those debts. Otherwise, unless their leadership is sufficiently armored, it will simply be decapitated.[50]

SETTLER COLONIALISM

Sovereignty connects to a different national question: settler colonization and decolonization within the First World, and the breaking of the social order in the settler-colonies of the Third World. This concerns land and environment, the two very tightly bound, since land alienation through settler-colonialism has had a very direct environmental impact. It has often meant the catastrophic and deliberate collapse of entire worlds.[51] As a political process, Kyle Powys Whyte argues, "Settler colonialism can be interpreted as a form of *environmental injustice* that wrongfully interferes with and erases the socioecological contexts required for indigenous populations to experience the world as a place infused with responsibilities to humans, nonhumans and ecosystems."[52] Anti-colonial struggle for control of the land is, then, a pathway to worldwide environmental restoration and renaissance. And in other settler states or politically decolonized settler states, nation demands concern first and foremost land. In Zimbabwe, for example, the most radical post-Cold War agrarian reform occurred, eliciting a "cold" campaign of demonization, isolation, and possibly, rollback.[53] In South Africa, agrarian reform is on the table, and is a vibrant political demand in cities, slums, and countryside alike.[54]

In the North American settler-states, Indigenous struggles at Standing Rock and through Idle No More have been catalysts for far broader consciousness amongst non-Indigenous radicals of the simmering and bubbling national question.[55] More broadly, it has been amongst the Indigenous that rights to access and use of the environment are intertwined with rights to land and Land Back. The Anchorage Declaration, which assembled Indigenous representatives from North America, the Arctic, Asia, Pacific, Latin America, Africa, and the Caribbean, called for states to "recognize and implement the fundamental human rights and status of Indigenous Peoples, including the collective rights to traditional ownership, use, access, occupancy, and title to traditional land, air, forests, waters, ocean, sea ice and sacred sites."[56] The Anchorage Declaration calls for restitution: to "return and restore lands, territories, waters," and "sacred sites that have been taken from Indigenous Peoples."[57] To call upon a state to do something demands that decolonization be a banner for action, including struggle from populations living in settler-colonial states to demand their governments restore land and treaty rights. And since land is the relationship underlying settler-colonial state formation, a relationship created by colonial primitive accumulation, restitution implies revolutionary change in world property structures. Similarly, the International Indigenous Peoples Forum on Climate Change connect such calls to their criticism of REDD and other carbon-offsetting and Clean Development Mechanism projects, noting that "land and resource rights" of Indigenous peoples need to be respected before any consideration of REDD or REDD+ carbon compensation.[58] That is, they recognize Indigenous peoples' need to be able to determine the political paths to decide how to structure their internal socio-ecologies. Similarly, the Red Nation demands:

> Treaty rights and Indigenous rights be applied and upheld both on- and off-reservation and federal trust land. All of North America, the Western Hemisphere, and the Pacific is Indigenous land. Our rights do not begin or end at imposed imperial borders we did not create nor give our consent to.[59]

Furthermore, the Red Nation explicitly draws the connections between politics and environmental management. They highlight how national self-determination is elemental and prior to acting ethically as custodians of nature: "We must first be afforded dignified lives as Native peoples who are free to perform our purposes as stewards of life if we are to protect and respect our nonhuman relatives – the land, the water, the air, the plants, and the animals."[60]

DECOLONIZATION AND SAFEGUARDING NATURE

In an under-remarked bit of historical poetic justice, biodiversity most flourishes in Indigenous-held parts of the planet. Rather than the idea that we should protect species by surrounding them with political or legal walls or placing them on preserves only accessible to the monied, non-human nature can be lived in and among.

Indigenous lands worldwide have equal to or higher biodiversity than "protected areas."[61] As the ecologist Victor Toledo points out, twelve countries – Brazil, Indonesia, Colombia, Australia, Mexico, Madagascar, Peru, China, Philippines, India, Ecuador, and Venezuela – have the highest number of species and endemic species, including mammals, birds, reptiles, amphibians, freshwater fishes, butterflies, tiger-beetles and flowering plants. Of those twelve countries, nine of them are on the list of 25 nations with the greatest number of endemic languages. Of the 233 marine, freshwater, and terrestrial bioregions with the fullest diversity of habitats and species, Indigenous peoples live in 136 of them. Half of the world's 3,000 Indigenous groups live in these eco-regions.[62] More recent work on Canada, Brazil, and Australia, three countries which have historically or continue to carry out genocide against the Indigenous, shows that biodiversity, from grizzly bears to kangaroos, frogs to songbirds, is highest and richest in Indigenous-managed lands.[63]

Protecting and respecting land treaty rights – Land Back – is the fast route to safeguarding the future. This is not because of some primordial and timeless capacity of Indigenous people to live in nature in peace and harmony. It is because Indigenous people are often engaged in primary production in the general sense. They have cosmologies based on a humane relationship to the land. And they carry out forms of production, from hunting to husbandry to horticulture, based on ancestral knowledge about how to live in, on, and with the land and to ride and work with and gently remold ecological cycles rather than chemically wrench and rework them with genetic tinkering and tampering and dousing plants in chemicals.[64]

This does not mean primary production should be the eternal lot of Indigenous people. As with every other people on Earth, they have rights to sovereign industrialization. It means Indigenous land management is superior to settler land management. Defense of Indigenous treaty rights and the restoration or restitution of land creates a safer world for all of humanity.[65] (In western Canada, for example, the members of the West Moberly and Saulteau are working to restore the caribou, a lesson in the merits of Indigenous landscape management and biodiversity conservation.[66])

In a bit of historical poetry, I write these words as apocalyptic wildfires have dyed the skies of the US West Coast a hazy orange. Indigenous techniques of land management, including controlled burns based on a holistic method of living in, rather than above or in control of nature, used to use fire to maintain "basketry materials, medicinal plants, acorn trees and hunting grounds," says Elizabeth Azzuz, a member of the Yurok Tribe. Before the settler invasion, large and frequent burns immolated fuel, which ensured that unintentional fires spread slowly, having no fuel on which to feed. The settler-capitalist invasion banned burning and built monoculture timber plantations in lieu of a tapestry of grasslands and agroforestry. In northwest California's Klamath basin, members of the Karuk and Yurok tribes have continued to carry out controlled burns, retaining their knowledge of fire science and transmitting that knowledge between generations. Jackie Fielder, a candidate for the California senate, and an Indigenous socialist, recently praised the Karuk efforts. A new revolutionary land-tenure regime, decolonization, and swift changes in land management practices drawing on rather than discarding Indigenous technical knowledge could avoid the next conflagration.[67] Land Back is neither surrender nor sacrifice, but the shift which makes the world big enough for all of us.

Conclusion

This book is about a People's Green New Deal. And not just any bundle of programs. A People's Green New Deal should be eco-socialist, and should explain what eco-socialism is, identify it as where we want to go, and think about how to get there. Eco-socialism is not green social democracy. Eco-socialism aspires to egalitarian redistribution, public ownership of the means of production, respectful, modest, and humane management of the human interaction with non-human nature, and decommodification of social life globally, not merely locally. And it aspires to eliminating capitalism not merely locally, but globally. And it is explicitly anti-imperialist, and pro-Third World state sovereignty, because states are necessary containers in our historical moment for Third World popular development, and for managing climate debt payments.

But getting to where we want to go means understanding properly where we are. The North and the South are not set up the same way, even though they are part of one world capitalist system. But even though they are different, they are not at different stages of development. The North is not more advanced than the South. The North is wealthy in its current form because the South is poor, and to keep the North wealthy under capitalism requires the South to remain poor and subject to imperialism. Furthermore, northern accumulation and development is qualitatively different from accumulation in the South. It relies on a broad-based domestic market, and its capitalists produce for that market. Southern accumulation rests much more on exporting its goods to the North. In this way, people in the South are generally more exploited than they are in the North. First, by super-exploitation. Second, by wholesale eradication and destruction of their societies, as is occurring in the Arab region, or has occurred under settler-colonialism, including in the United States. Third, by primitive accumulation, including use of atmospheric space, giving rise to climate debts. And fourth, by appropriation of their resources on a large scale.

Eco-socialist strategy has to contend with these fundamental differences. That does not mean northern or southern perspectives are in a zero-sum game any more than gender oppression suggests women and men are in a zero-sum game or Black and white populations are in zero-sum games.

Rather, it suggests that to have inclusive socialist projects, the perspectives and demands of those most oppressed and dispossessed have to be the foundation of analysis and strategy.

Yet, this brings us to a fundamental problem in Global North climate politics, environmental politics, and politics more broadly. Being has a tendency to determine consciousness. Northern beings tend to be less oppressed than southern beings, because northern societies are far wealthier on average, and also have substantial middle classes accustomed to an imperial mode of living. For that reason, it is an Olympian task to detangle northern social democracy from imperialism. As a result, many choose to avoid the difficult and focus on the pragmatic. Pragmatism appears as a recurring tendency in northern climate politics and left politics more broadly to not take on the anti-imperialist burden of transformation. They kick climate debt payments down the road. And anti-imperialism – committed opposition to Western wars and sanctions – is weak or non-existent in contemporary left climate politics.

This tableau is not inevitable. The arc of struggle extending from the Battle of Seattle to the opposition to the War on Iraq to the Cochabamba People's Agreements shows that North and South can move towards one another ideologically, organizationally, and politically. A committed anti-imperialism and internationalism emerges from humanism and decency, radicalized in a way which incorporates the needs of the planet and everyone on it. North–South distance is produced historically by people in struggle or out of struggle or in surrender. It can also be un-made by people in struggle, as long as we do not surrender.

But how? It is increasingly common to argue that social democracy is so distant a horizon that getting there would be good enough, or from another perspective, that the insurgent Sanders, Corbyn, and Mélenchon efforts prove social democracy is practically peeking over the horizon. Being so close, why not throw dozens of self-styled socialist candidates in office, press through a populist social democratic Green New Deal, and then fight capitalism and imperialism a bit later? Such a call has a huge understandable appeal in the North, when it and especially the United States see people in prison, on the streets, going hungry, and dying in huge numbers from a failed medical system even before the pandemic. The core states are not humane societies even for their own citizens.

It is true that the situation is bad and that efforts to change it are enormously alluring – although it is not true at all that social democracy was in any way imminent merely due to the understandable popularity of Sanders or other kindred candidates and movements, across Europe and the United States. Indeed, the absolute devastation, failure, and defeat of one such movement after another, from SYRIZA to Podemos to Corbyn to Sanders, should be taken as a warning. That is why I explained the historical circumstances across the core which gave birth to social democracy, or the 1940s–1960s social welfare programs. Most writing on those topics attributes the reason to their success to labor movements, peoples' movements, and opaquely, "the Cold War." The Cold War means simply that 1917 and especially the success of the USSR in smashing the Nazi armies gave Communism incredible legitimacy in South and North alike. To avoid Communism in the South, the North encouraged the "development project" and coldly tolerated national-popular leaders as a domestic buffer against the left, at least until it dealt with the leftists via the Jakarta Method of mass murder, and coup d'état against national-popular leadership.[1] To avoid Communism in the North, political leaders were elected on reformist programs to emplace social-welfare states of every kind. There was little other option to prevent the red shadow from falling across Europe and elsewhere.

We can take three points from this. One: popular development is the outcome of class struggle. It is an achievement from below, not a grant from above. Two: class struggle is global, even if the lines of causality are hard to see. Three: global class struggle historically included those fighting for the abolition, not the softening, of capitalism.

What does that imply for the present? In essence, the struggle for domestic green social democracy as a transitional stage, and shorn of genuine anti-imperialist internationalism, is doomed to fail. No popular development of any kind has ever occurred in the North or South without a fight for something far more radical on the home front and abroad. On the other hand, if one builds a struggle around a theory and ideology that takes as its end point a foreshortened horizon, one will not get even to such a compromised target. Even avowedly revolutionary parties which come to take direct state power do not achieve their goals. If far more reformist-oriented parties or individuals occupy portions of the parliament, or even the executive of the state itself, they will be prevented even from reaching less radical horizons. This is the nature of change in a capitalist world. We cannot trick capitalism and we cannot legislatively gnaw away at accumulation circuits in the purely parliamentary fashion currently in vogue.

Furthermore, those who are not included and their programs accepted as non-negotiables will suffer the costs of northern reformism and opportunism. If capital is under pressure from social movements or political parties and needs to find a way to give something away to domestic middle classes, that something has to come from somewhere. And that somewhere will be those not included in the social and political struggles: in other words, the South. This warning is not a hypothetical. The exclusion of climate debt repayment from northern Green New Deal talk already shows that some northern leftists are preparing programs that will exclude the most fundamental and unified demands from the South: the People's Agreement of Cochabamba. The Sanders's plan's heavy emphasis on raw materials extraction, the under-emphasis on climate debt repayments, and the continued commitment to the military budget means that such a plan – anyway, now a dead letter – was never aiming to transform the world system. Its emphasis on heavy extraction of minerals meant the South would suffer the costs of a northern clean transition. So, what is the alternative?

TOWARDS A MOVEMENT OF MOVEMENTS?

For a People's Green New Deal to be planetary and eco-socialist, it has to build anti-imperialism and internationalism in from the outset. The difficulty for this process lies in how much northern capitalist and northern consumerism rest on South–North value flows: capital and the claims to plains and savannahs and pine forests, to cheap strawberries and apples and sugarcane and soy beans, manganese, lithium, copper, and oil, water and privileges to pollute, and the super-exploited labor of picking fruits and vegetables, mining minerals, and crafting iPhones – the myriad things for which northern wealth exchanges.

Fighting amidst, amongst, and against the continuation of those capital flows can take economic, productive, and political forms. In fact, they need to be combined. The more production can occur in the North, without relying on extracting and despoiling land, labor, and life from the South, the more the productive basis of the North can be rebuilt in a way which complements that of the South, and the more the North can free up resources for the settlement of climate debts. At the same time, in focusing on reworking northern productive sectors, we will see the glimmers of the political agent for just transition, based on building from existing strengths, getting labor on board for the green shift, and showing people how life can be made better even while combating imperialism. The more we can show that such shifts can be combined with overt internationalism in the form of climate debt payments and respect for peripheral political sovereignty, the more it will be feasible to forge the necessary alliances and social bases for a mass-based internationalist project of permanent social change.

Such a transition does not imply a reduction in northern quality of life. Indeed, the current social system is based on massive looting from the South, and northern quality of life is horrible. Instead, it means removing some sectors and growing others. And it means controlling industrialization, for the sake of humanity, the planet, and the South. To control industrialization does not mean to eliminate industrialization, let alone modern social life with complex forms of economic interchange and interdependence. It means understanding how on the one hand, the North is gratuitously over-industrialized, and not to the benefit of working-class life. And it means accepting how much northern industrial capital, and the consumption which it encourages, rests on de-development or underdevelopment of the South. Industry is part of a global process, where more advanced goods concentrate in the core and waste, pollution, and poverty concentrate in the periphery. Much of this is what the degrowth conversation refers to, and we all can agree that there are sectors of the core economies which should be vastly retooled or eliminated.

An analytical foundation stone of this approach is that capitalism is not a system of production of useful things, but a system for the production of waste.[2] Under capitalism, people die before they should given the possibilities of existing technologies and productive forces, and they die before they should because of which technologies are emphasized and how they are distributed. Eventually capitalism fetters the productive forces. Eliminating capitalism means eliminating the myriad inefficiencies and irrationalities which exist because of a system of production for profit. Changing the kinds of things we have and produce, from agriculture to manufacturing and industrialization to means of transportation to healthcare to education would mean producing more things useful for humanity, using fewer resources, and reducing waste to what we are willing and able to spend the resources to clean up and remediate.

DEVELOPMENT BY POPULAR PROTECTION

To begin to carry out popular development in the North in a way that makes people's lives in the North better while also making people's lives in the South better, we need to look towards sectors and industries where we can increase the production of things which are useful to poor people without causing further social or environmental harm to the South. There are lots of easy targets, those many things only needed in order to protect capitalism's capacity to produce for profit. First of all, we should eliminate the use of northern productive capacity to produce pure waste: the military. With that huge logistical and material sinkhole closed up, it will be much easier to devote northern productive resources to northern human needs.

Using solely domestic sources of renewable energy and building the machinery domestically would be preferable to emerging forms of green-energy colonialism, biofuel land-grabbing, nuclear, or oil and coal. And the inputs from such machinery should as much as possible be sourced locally (to the extent they are not, they become subject to price negotiations on the international market. There are ways to make those fairer, as well). Labor and production converge, since this would create more jobs for unemployed or underemployed people in the core, while ensuring that those jobs' material basis is not looting from the South via primitive accumulation or unequal exchange.

Shifting to less industrial and more knowledge-centered forms of healthcare would give people in the core more jobs, for training healthcare workers and for healthcare workers themselves. We could achieve much better healthcare outcomes with less use of industrialized and reactive medicine, focusing instead on community-centered preventative medicine focused on primary care, including nutrition. This is without even getting into what the modern medical research system could look like if it were focusing on goals like near-zero environmental impact and maximizing human well-being on a global level, rather than how it is generally organized now: for profit.

Agriculture is central, precisely because it can be done without using any material or labor from the periphery. And in so doing, using agroecology, we can protect the natural environment and biodiversity, take care of the land, and ward off or build buffers against viruses that are used to further impoverish the poor on a global scale.

And who is to do what is to be done? Community energy coops cascading countrywide; doctors, nurses, and common people agitating for national preventative and free healthcare; displaced and often illegalized forced-migrant and refugee communities at the beating heart of Third World solidarity; Leninist and autonomist formations focused on serving the people; research clusters and agronomic research institutions rebuilding the intellectual foundation for responsible landscape management; ecological populist designers and architects, artists and artisans working with appropriate materials; municipal efforts to graft cellulosic CO_2 absorbing greenery onto gray spaces; and village, town, and city-level popular referenda for massive investments in public transportation and shifts to biking and walking – human-scale cities. All of this can be done with an internationalist perspective, while demanding climate debts be paid and national sovereignty in the Third World respected and settler-states decolonized. Will it be done? That is a matter of politics, which is to say it is a matter of struggle and choice.

Notes

INTRODUCTION

1. Cited in Robinson Meyer, "So Has the Green New Deal Won Yet?," *The Atlantic*, November 15, 2019, www.theatlantic.com/science/archive/2019/11/did-green-new-deal-win-look-after-one-year/602032/.
2. Stephen R. Gliessman, *Agroecología: procesos ecológicos en agricultura sostenible* (CATIE, 2002), 13.
3. Thomas L. Friedman, "Thomas L. Friedman: The Power of Green," *The New York Times*, April 15, 2007, sec. Opinion, www.nytimes.com/2007/04/15/opinion/15iht-web-0415edgreen-full.5291830.html.
4. "A Green New Deal for Europe," Election platform, 2009, https://dnpprepo.ub.rug.nl/562/.
5. Charlotte Kates, "Criminalizing Resistance," *Jacobin* (blog), January 27, 2014, www.jacobinmag.com/2014/01/criminalizing-resistance/.
6. Alexandria Ocasio-Cortez, "Text – H.Res.109 – 116th Congress (2019–2020): Recognizing the Duty of the Federal Government to Create a Green New Deal," webpage, February 12, 2019, 2019/2020, www.congress.gov/bill/116th-congress/house-resolution/109/text.
7. Ocasio-Cortez.
8. "DSA's Green New Deal Principles – DSA Ecosocialists," accessed February 1, 2021, https://ecosocialists.dsausa.org/2019/02/28/gnd-principles/.
9. Zak Cope, *The Wealth of (Some) Nations: Imperialism and the Mechanics of Value Transfer* (Pluto Press, 2019).
10. "Call:," *World People's Conference on Climate Change and the Rights of Mother Earth* (blog), January 15, 2010, https://pwccc.wordpress.com/2010/01/15/call/.
11. Naomi Klein, "A New Climate Movement in Bolivia," *The Nation*, accessed August 20, 2020, www.thenation.com/article/archive/new-climate-movement-bolivia/.
12. "Rights of Mother Earth," *World People's Conference on Climate Change and the Rights of Mother Earth* (blog), January 4, 2010, https://pwccc.wordpress.com/programa/.
13. "People's Agreement of Cochabamba," *World People's Conference on Climate Change and the Rights of Mother Earth* (blog), April 24, 2010, https://pwccc.wordpress.com/2010/04/24/peoples-agreement/. These are all direct quotations from the document.

14. Caspar A. Hallmann et al., "More than 75 Percent Decline over 27 Years in Total Flying Insect Biomass in Protected Areas," *PLOS ONE* 12, no. 10 (October 18, 2017): e0185809, https://doi.org/10.1371/journal.pone.0185809.
15. Sam Moyo, Praveen Jha, and Paris Yeros, "The Classical Agrarian Question: Myth, Reality and Relevance Today," *Agrarian South: Journal of Political Economy* 2, no. 1 (2013): 113.
16. Samir Amin, "Accumulation and Development: A Theoretical Model," *Review of African Political Economy*, no. 1 (August 1, 1974): 9, https://doi.org/10.2307/3997857.
17. Samir Amin, *Accumulation on a World Scale: A Critique of the Theory of Underdevelopment* (Monthly Review Press, 1974); Giovanni Arrighi and Jessica Drangel, "The Stratification of the World-Economy: An Exploration of the Semiperipheral Zone," *Review (Fernand Braudel Center)* 10, no. 1 (1986): 9–74.
18. Frantz Fanon, *The Wretched of the Earth*, tr. from the French by Constance Farrington (Grove Press, 1963), 102.
19. Walter Rodney, *How Europe Underdeveloped Africa* (Fahamu/Pambazuka, 2012); Eduardo Galeano, *Open Veins of Latin America: Five Centuries of the Pillage of a Continent* (NYU Press, 1997).
20. Frantz Fanon, *The Wretched of the Earth*, tr. from the French by Constance Farrington (Grove Press, 1963), 102.

CHAPTER 1

1. Jonathan Woetzel et al., "Climate Risk and Response, Physical Hazards and Socioeconomic Impacts" (McKinsey Consulting, January 2020).
2. Robert W. McChesney, *The Political Economy of Media: Enduring Issues, Emerging Dilemmas* (Monthly Review Press, 2008).
3. "Nature Risk Rising: Why the Crisis Engulfing Nature Matters for Business and the Economy," World Economic Forum, 8, accessed October 17, 2020, www.weforum.org/reports/nature-risk-rising-why-the-crisis-engulfing-nature-matters-for-business-and-the-economy/.
4. Philip McMichael, "Commentary: Food Regime for Thought," *The Journal of Peasant Studies* 43, no. 3 (2016): 660.
5. Daniela Gabor, "The Wall Street Consensus," July 2020, https://osf.io/preprints/socarxiv/wab8m/.
6. Angela Mitropoulos, "Playing With Fire: Securing the Borders of a Green New Deal," *New Socialist*, January 12, 2020, http://newsocialist.org.uk/playing-fire-securing-borders/.
7. Walter Benjamin, "On the Concept of History," 1940, www.marxists.org/reference/archive/benjamin/1940/history.htm.
8. "Nature Risk Rising," 20–21.
9. Giovanni Batz, "Destabilized by US Imperialism, Central America Now Faces Climate Catastrophes," *The Red Nation* (blog), November 20, 2020, http://

theirednation.org/destabilized-by-us-imperialism-central-america-now-faces-climate-catastrophes/.

10. Andres Schipani, "Bolivia's 'Environmentalist' President Morales under Fire over Amazon," August 30, 2019, www.ft.com/content/a7bd5aea-c92d-11e9-a1f4-3669401ba76f; For context disputing any claims of ecological irresponsibility, see "Federico Fuentes on Twitter," Twitter, accessed October 30, 2020, https://twitter.com/FredFuentesGLW/status/1169516326227808256.
11. Nick Estes, *Our History Is the Future: Standing Rock Versus the Dakota Access Pipeline, and the Long Tradition of Indigenous Resistance* (Verso Books, 2019); Roxanne Dunbar-Ortiz, *An Indigenous Peoples' History of the United States* (Beacon Press, 2014).
12. Arun Kundnani, *The Muslims Are Coming!: Islamophobia, Extremism, and the Domestic War on Terror* (Verso Books, 2014).
13. Amar Bhattacharya et al., "Aligning G20 Infrastructure Investment with Climate Goals & the 2030 Agenda," Foundations 20 Platform, a Report to the G20 (The Brookings Institution & Global Development Policy Center, 2019).
14. "Future We Want – Outcome Document. Sustainable Development Knowledge Platform," accessed October 18, 2020, https://sustainabledevelopment.un.org/futurewewant.html.
15. Bridget O'Laughlin, "Book Reviews," *Development and Change* 35, no. 2 (2004): 387, https://doi.org/10.1111/j.1467-7660.2004.00357.x.
16. David Spratt and Ian Dunlop, "What Lies Beneath: The Understatement of Existential Climate Risk," *Breakthrough (National Centre for Climate Restoration)*, 2018.
17. Martin Glaberman, *Wartime Strikes: The Struggle Against the NoStrike Pledge in the UAW During World War II* (Bewick/Ed, 1980); Bruce Nelson, "Organized Labor and the Struggle for Black Equality in Mobile during World War II," *The Journal of American History* 80, no. 3 (1993): 952–88.
18. John Bellamy Foster, "Malthus' Essay on Population at Age 200: A Marxian View," *Monthly Review* 50, no. 7 (1998): 1.
19. Amartya Sen, "Ingredients of Famine Analysis: Availability and Entitlements," *The Quarterly Journal of Economics* 96, no. 3 (August 1, 1981): 433–64, https://doi.org/10.2307/1882681.
20. Utsa Patnaik, "Profit Inflation, Keynes and the Holocaust in Bengal, 1943–44," *Economic & Political Weekly* 53, no. 42 (2018): 33; Mike Davis, *Late Victorian Holocausts: El Nino Famines and the Making of the Third World* (Verso Books, 2002).
21. Frances Moore Lappé et al., *World Hunger: 12 Myths* (Grove Press, 1998).
22. William J. Ripple et al., "World Scientists' Warning of a Climate Emergency," *BioScience*, accessed December 2, 2019, https://doi.org/10.1093/biosci/biz088.
23. "Our Patrons," Population Matters | Every Choice Counts | Sustainable World Population, September 19, 2018, https://populationmatters.org/our-patrons.
24. Betsy Hartmann, "Converging on Disaster: Climate Security and the Malthusian Anticipatory Regime for Africa," *Geopolitics* 19, no. 4 (2014):

757–83; Diana Ojeda, Jade S. Sasser, and Elizabeth Lunstrum, "Malthus's Specter and the Anthropocene," *Gender, Place & Culture* 27, no. 3 (2020): 316–32.

25. Amanda Shaw and Kalpana Wilson, "The Bill and Melinda Gates Foundation and the Necro-Populationism of 'Climate-Smart' Agriculture," *Gender, Place & Culture* 27, no. 3 (August 7, 2019): 370–93, https://doi.org/10.1080/0966369X.2019.1609426.
26. Ruy Mauro Marini, "Subdesarrollo y Revolución" (Siglo Veintiuno Editores Mexico City, 1969).
27. Ali Kadri, *Arab Development Denied: Dynamics of Accumulation by Wars of Encroachment* (Anthem Press, 2015); Ali Kadri, *Imperialism with Reference to Syria* (Springer, 2019).
28. Benjamin Neimark, Oliver Belcher, and Patrick Bigger, "US Military is a Bigger Polluter than as Many as 140 Countries – Shrinking This War Machine Is a Must," *The Conversation*, accessed January 13, 2021, http://theconversation.com/us-military-is-a-bigger-polluter-than-as-many-as-140-countries-shrinking-this-war-machine-is-a-must-119269.
29. "Plan for Climate Change and Environmental Justice | Joe Biden," Joe Biden for President: Official Campaign Website, accessed November 16, 2020, https://joebiden.com/climate-plan/.
30. Department of Defense, "National Security Implications of Climate-Related Risks and a Changing Climate" (Department of Defense, July 23, 2015), 3–4.
31. Office of the Under Secretary of Defense for Acquisition and Sustainment, "Report on Effects of a Changing Climate to the Department of Defense," January 2019, 5.
32. Office of the Under Secretary of Defense for Acquisition and Sustainment, 10–13.
33. https://medium.com/@teamwarren/our-military-can-help-lead-the-fight-in-combating-climate-change-2955003555a3.
34. "Department of Defense Climate Resiliency and Readiness Act (S. 1498)," GovTrack.us, accessed February 19, 2020, www.govtrack.us/congress/bills/116/s1498.
35. "Department of Defense Climate Resiliency and Readiness Act (S. 1498)," 23–24.
36. "Department of Defense Climate Resiliency and Readiness Act (S. 1498)," 26–27.
37. Jeremy Rifkin, *The Green New Deal: Why the Fossil Fuel Civilization Will Collapse by 2028, and the Bold Economic Plan to Save Life on Earth* (St Martin's Publishing Group, 2019), 132–39.
38. Bhattacharya et al., "Aligning G20 Infrastructure Investment with Climate Goals & the 2030 Agenda," 3.
39. Werner Hoyer, "Without Private Finance, There Will Be No Green Transition. Here Is What Needs to Happen," World Economic Forum, accessed February 19, 2020, https://www.weforum.org/agenda/2019/09/how-we-should-be-investing-in-the-green-transition/.

40. Climate Finance Leadership Initiative, "Financing the Low-Carbon Future: A Private-Sector View on Mobilizing Climate Finance," September 2019, 8.
41. Climate Finance Leadership Initiative, 10.
42. Climate Finance Leadership Initiative, 47.
43. "Technology Transfer Will be Part of Copenhagen Climate Deal," *Intellectual Property Watch* (blog), September 16, 2009, www.ip-watch.org/2009/09/16/technology-transfer-will-be-part-of-copenhagen-climate-deal/.
44. Climate Finance Leadership Initiative, 11.
45. Rifkin, *The Green New Deal*, 24.
46. Rifkin, 141.
47. Rifkin, 185.
48. Caroline Lucas, "A Green New Deal Offers Hope for a Better Future – We Need to Set out a Positive Vision," *The Green New Deal Group* (blog), October 22, 2019, https://greennewdealgroup.org/a-green-new-deal-offers-hope-for-a-better-future-we-need-to-set-out-a-positive-vision/.
49. Mariana Mazzucato, "How Industrial Strategy Can Drive a Green New Deal," IPPR, October 23, 2019, www.ippr.org/blog/industrial-strategy-drive-green-new-deal-mariana-mazzucato.
50. Gabor, "The Wall Street Consensus," 20.
51. Edward O. Wilson, *Half-Earth: Our Planet's Fight for Life* (W. W. Norton & Company, 2016).
52. https://allianceforscience.cornell.edu/blog/2018/08/sparing-half-earth-nature-still-feeding-humanity/. https://allianceforscience.cornell.edu/about/funders/.
53. Bram Büscher, "Reassessing Fortress Conservation? New Media and the Politics of Distinction in Kruger National Park," *Annals of the Association Of American Geographers* 106, no. 1 (2016): 114–29, https://doi.org/10.1080/00045608.2015.1095061.
54. A. Deutz and G. Heal, "Financing Nature: Closing the Global Biodiversity Finance Gap Report," The Paulson Institute, The Nature Conservancy and the Cornell Atkinson Center for Sustainability, 2020, www.nature.org/en-us/what-we-do/our-insights/reports/financing-nature-biodiversity-report/.
55. John Lynch and Raymond Pierrehumbert, "Climate Impacts of Cultured Meat and Beef Cattle," *Frontiers in Sustainable Food Systems* 3 (2019), https://doi.org/10.3389/fsufs.2019.00005.
56. "Stordalen Foundation," Stordalen Foundation, accessed October 18, 2020, www.stordalenfoundation.no; "What Is EAT," EAT, accessed October 18, 2020, https://eatforum.org/about/who-we-are/what-is-eat/.
57. Diana K. Davis, *Resurrecting the Granary of Rome: Environmental History and French Colonial Expansion in North Africa* (Ohio University Press, 2007).
58. "Intensive Monoculture Is Putting Water Systems in Peril," accessed January 24, 2021, https://phys.org/news/2020-09-intensive-monoculture-peril.html.
59. Vera, *Grazing Ecology and Forest History*.
60. Prafulla Kalokar and Kanna Siripurapu, "Is the Environment for Taking From or for Giving To? A Young Indigenous Economist Finds Answers in His

Own Culture," *Terralingua* (blog), October 16, 2020, https://terralingua.org/langscape_articles/is-the-environment-for-taking-from-or-for-giving-to-a-young-indigenous-economist-finds-answers-in-his-own-culture/.

61. "Stordalen Foundation," Stordalen Foundation, accessed October 18, 2020, www.stordalenfoundation.no; "What Is EAT," EAT, accessed October 18, 2020, https://eatforum.org/about/who-we-are/what-is-eat/.
62. https://www.wbcsd.org/Overview/About-us.
63. The Food and Land Use Coalition, "Growing Better:Ten Critical Transitions to Transform Food and Land Use," September 2019, 122.
64. Walter Willett et al., "Food in the Anthropocene: The EAT–Lancet Commission on Healthy Diets from Sustainable Food Systems," *The Lancet* 393, no. 10170 (February 2, 2019): 447–92, https://doi.org/10.1016/S0140-6736(18)31788-4.
65. Wim Carton, "Carbon Unicorns and Fossil Futures. Whose Emission Reduction Pathways is the IPCC Performing?," 2020.
66. "Australian Industry Energy Transitions Initiative | ETC," *Energy Transitions Commission* (blog), accessed October 7, 2020, www.energy-transitions.org/publications/australian-industry-energy-transitions-initiative/.
67. European Commission, "Powering a Climate-Neutral Economy: An EU Strategy for Energy System Integration" (European Union, July 2020).
68. Senate Democrats' Special Committee on the Climate Crisis, "The Case for Climate Action," August 25 2020, www.schatz.senate.gov/imo/media/doc/SCCC_Climate_Crisis_Report.pdf.
69. Energy Transitions Commission, "Towards a Low-Carbon Steel Sector," *Energy Transitions Commission* (blog), 2019, www.energy-transitions.org/publications/towards-a-low-carbon-steel-sector/.
70. Corporate Europe Observatory, "Research and Destroy: The Factories of the Industrial Bioeconomy Threaten the Climate and Biodiversity," 2020.
71. "Making Mission Possible: Delivering a Net-Zero Economy" (Energy Transitions Commission, 2020), www.energy-transitions.org/publications/making-mission-possible/.
72. Thibaud Clisson, "Is Carbon Capture and Storage a Complete Waste of Time and Effort?," *Investors' Corner, BNP Paribas* (blog), May 29, 2018, https://investors-corner.bnpparibas-am.com/investing/carbon-capture-storage/.

CHAPTER 2

1. Arnim Scheidel et al., "Ecological Distribution Conflicts as Forces for Sustainability: An Overview and Conceptual Framework," *Sustainability Science* 13, no. 3 (2018): 585–98.
2. Thanks to Zeyad El Nabolsy for suggesting ways to clarify my argument here.
3. David F. Noble, "Social Choice in Machine Design: The Case of Automatically Controlled Machine Tools, and a Challenge for Labor," *Politics & Society* 8, no. 3–4 (1978): 313–47; Andreas Malm, *Fossil Capital: The Rise of Steam-Power and the Roots of Global Warming* (Verso, 2016).

4. Amiya Kumar Bagchi, *Perilous Passage: Mankind and the Global Ascendancy of Capital* (Rowman & Littlefield Publishers, 2008).
5. Mike Davis, *Late Victorian Holocausts: El Nino Famines and the Making of the Third World* (Verso Books, 2002).
6. Fernando Estenssoro and Eduardo Deves Valdés, "Antecedentes históricos del debate ambiental global: Los primeros aportes latinoamericanos al origen del concepto de Medio Ambiente y Desarrollo (1970–1980)," *Estudos Ibero-Americanos* 39, no. 2 (2013): 237–61; Max Ajl, "Delinking's Ecological Turn: The Hidden Legacy of Samir Amin," ed. Ushehwedu Kufakurinani, Ingrid Harvold Kvangraven, and Maria Dyveke Styve, *Review of African Political Economy*, no. Samir Amin and Beyond: Development, Dependence and Delinking in the Contemporary World (2021).
7. Clifford D. Conner, *A People's History of Science: Miners, Midwives, and Low Mechanicks* (Hachette UK, 2009).
8. Mark Elvin, "Why China Failed to Create an Endogenous Industrial Capitalism," *Theory and Society* 13, no. 3 (1984): 379–91.
9. "People's Agreement of Cochabamba"; Rikard Warlenius, Gregory Pierce, and Vasna Ramasar, "Reversing the Arrow of Arrears: The Concept of 'Ecological Debt' and Its Value for Environmental Justice," *Global Environmental Change* 30 (2015): 21–30.
10. Utsa Patnaik, "Revisiting the 'Drain', or Transfers from India to Britain in the Context of Global Diffusion of Capitalism," in *Agrarian and Other Histories: Essays for Binay Bhushan Chaudhuri*, ed. Shubhra Chakrabarti and Utsa Patnaik (Tulika Books, 2017), 277–318; Alec Gordon, "Netherlands East Indies: The Large Colonial Surplus of Indonesia, 1878–1939," *Journal of Contemporary Asia* 40, no. 3 (August 1, 2010): 425–43, https://doi.org/10.1080/00472331003798392; Sidney Wilfred Mintz, *Sweetness and Power: The Place of Sugar in Modern History* (Penguin Books, 1986); Alf Hornborg, "Footprints in the Cotton Fields: The Industrial Revolution as Time–Space Appropriation and Environmental Load Displacement," *Ecological Economics* 59, no. 1 (2006): 74–81; Eric Williams, *Capitalism and Slavery* (UNC Press Books, 2014).
11. Rodney, *How Europe Underdeveloped Africa*; Amin, *Accumulation on a World Scale*; Celso Furtado, "Development and Stagnation in Latin America: A Structuralist Approach," *Studies in Comparative International Development* 1, no. 11 (November 1965): 159–75, https://doi.org/10.1007/BF02800594; Vania Bambirra, *El Capitalismo Dependiente Latinoamericano* (Siglo XXI, 1999).
12. Chad Montrie, *A People's History of Environmentalism in the United States* (A&C Black, 2011), 119–30.
13. Max Ajl, "Auto-Centered Development and Indigenous Technics: Slaheddine El-Amami and Tunisian Delinking," *Journal of Peasant Studies* 46, no. 6 (2019): 1240–63; Max Ajl and Divya Sharma, "Transversal Countermovements: The Afterlives of the Green Revolution in Tunisia and India." (15th Meeting of International Society for Ecological Economics, Puebla, Mexico, September 10, 2018); Artemio Cruz León et al., "La Obra Escrita de Efraím Hernández

Xolocotzi, Patrimonio y Legado," *Revista de Geografía Agrícola*, no. 50–51 (2013): 7–29; Vandana Shiva, *The Violence of the Green Revolution: Third World Agriculture, Ecology, and Politics* (University Press of Kentucky, 2016); Víctor M. Toledo and Narciso Barrera-Bassols, *La Memoria Biocultural: La Importancia Ecológica de Las Sabidurías Tradicionales*, vol. 3 (Icaria editorial, 2008); Azzam Mahjoub, "Technologie et Développement" (Universite d'Aix-Marseille-II, 1982).

14. Fernando Estenssoro and Eduardo Deves Valdés, "Antecedentes históricos del debate ambiental global: Los primeros aportes latinoamericanos al origen del concepto de Medio Ambiente y Desarrollo (1970–1980)," *Estudos Ibero-Americanos* 39, no. 2 (2013): 237–61; Max Ajl, "Delinking's Ecological Turn: The Hidden Legacy of Samir Amin," ed. Ushehwedu Kufakurinani, Ingrid Harvold Kvangraven, and Maria Dyveke Styve, *Review of African Political Economy*, no. Samir Amin and Beyond: Development, Dependence and Delinking in the Contemporary World (2021).
15. Arthur P. J. Mol, "Ecological Modernisation and Institutional Reflexivity: Environmental Reform in the Late Modern Age," *Environmental Politics* 5, no. 2 (June 1, 1996): 310, https://doi.org/10.1080/09644019608414266.
16. Maarten Hajer, "Ecological Modernisation as Cultural Politics," *Risk, Environment and Modernity: Towards a New Ecology* 253 (1996).
17. John Bellamy Foster, "The Planetary Rift and the New Human Exemptionalism: A Political-Economic Critique of Ecological Modernization Theory," *Organization & Environment* 25, no. 3 (2012): 228.
18. Michael Shellenberger and Ted Nordhaus, "The Death of Environmentalism," *Geopolitics, History, and International Relations* 1, no. 1 (2009): 121–63.
19. "An Ecomodernist Manifesto," accessed February 10, 2020, www.ecomodernism.org.
20. "An Ecomodernist Manifesto," 6–7.
21. "An Ecomodernist Manifesto," 7, 9.
22. "An Ecomodernist Manifesto," 12.
23. "An Ecomodernist Manifesto," 17.
24. "An Ecomodernist Manifesto," 20.
25. "An Ecomodernist Manifesto," 29.
26. Carolyn Merchant, *The Death of Nature: Women, Ecology, and the Scientific Revolution* (HarperCollins, 1990).
27. Dina Gilio-Whitaker, *As Long As Grass Grows: The Indigenous Fight for Environmental Justice from Colonization to Standing Rock* (Beacon Press, 2019).
28. Aaron Bastani, *Fully Automated Luxury Communism: A Manifesto* (Verso Books, 2020), 117–31.
29. Cited in "Asteroid Mining to Shape the Future of Our Wealth," *Interesting Engineering*, November 6, 2020, https://interestingengineering.com/asteroid-mining-to-shape-the-future-of-our-wealth.
30. Marc M. Cohen, "Robotic Asteroid Prospector (RAP)," n.d., 86.

31. Aaron Bastani, *Fully Automated Luxury Communism: A Manifesto* (Verso Books, 2020), 165–67.
32. Harry M. Cleaver, "The Contradictions of the Green Revolution," *The American Economic Review* 62, no. 1/2 (1972): 177–86; Raj Patel, "The Long Green Revolution," *The Journal of Peasant Studies* 40, no. 1 (2013): 1–63; Divya Sharma, "Techno-Politics, Agrarian Work and Resistance in Post-Green Revolution Punjab, India" (Dissertation, Cornell University, 2017).
33. Cited in Sigrid Schmalzer, *Red Revolution, Green Revolution: Scientific Farming in Socialist China* (University of Chicago Press, 2016), 2.
34. Ann Raeboline Lincy Eliazer Nelson, Kavitha Ravichandran, and Usha Antony, "The Impact of the Green Revolution on Indigenous Crops of India," *Journal of Ethnic Foods* 6, no. 1 (October 1, 2019): 8, https://doi.org/10.1186/s42779-019-0011-9.
35. Richa Kumar, "India's Green Revolution and Beyond," *Economic and Political Weekly* 54, no. 34 (2019): 41.
36. Max Ajl and Divya Sharma, "Transversal Countermovements: The Afterlives of the Green Revolution in Tunisia and India." (15th Meeting of International Society for Ecological Economics, Puebla, Mexico, September 10, 2018).
37. Aaron Bastani, *Fully Automated Luxury Communism: A Manifesto* (Verso Books, 2020), 169.
38. "Nutrition Country Profiles: Bangladesh Summary," accessed January 31, 2021, www.fao.org/ag/agn/nutrition/bgd_en.stm.
39. Aaron Bastani, *Fully Automated Luxury Communism: A Manifesto* (Verso Books, 2020), 169–70.
40. "U.S. Could Feed 800 Million People with Grain That Livestock Eat, Cornell Ecologist Advises Animal Scientists," *Cornell Chronicle*, August 7, 1997, https://news.cornell.edu/stories/1997/08/us-could-feed-800-million-people-grain-livestock-eat.
41. Hannah Ritchie and Max Roser, "Land Use," *Our World in Data*, November 13, 2013, https://ourworldindata.org/land-use.
42. Aaron Bastani, *Fully Automated Luxury Communism: A Manifesto* (Verso Books, 2020), 170.
43. Carolyn S. Mattick et al., "Anticipatory Life Cycle Analysis of In Vitro Biomass Cultivation for Cultured Meat Production in the United States," *Environmental Science & Technology* 49, no. 19 (October 6, 2015): 11941–49, https://doi.org/10.1021/acs.est.5b01614.
44. Jason Hickel and Giorgos Kallis, "Is Green Growth Possible?," *New Political Economy*, 2019, 1–18.
45. Xujia Jiang et al., "Revealing the Hidden Health Costs Embodied in Chinese Exports," *Environmental Science & Technology* 49, no. 7 (April 7, 2015): 4381–88, https://doi.org/10.1021/es506121s.
46. T. Wiedmann and M. Lenzen, "Environmental and Social Footprints of International Trade," *Nature Geoscience* 11 (2018), 314–21.

47. Alexander Dunlap, "Counterinsurgency for Wind Energy: The Bíi Hioxo Wind Park in Juchitán, Mexico," *The Journal of Peasant Studies* 45, no. 3 (March 19, 2018): 630–52, https://doi.org/10.1080/03066150.2016.1259221.
48. Giovanni Arrighi, Beverly J. Silver, and Benjamin D. Brewer, "Industrial Convergence, Globalization, and the Persistence of the North–South Divide," *Studies in Comparative International Development* 38, no. 1 (2003): 3–31.
49. IPCC, "Climate Change and Land: Summary for Policymakers," August 2019, www.ipcc.ch/site/assets/uploads/2019/08/4.-SPM_Approved_Microsite_FINAL.pdf.
50. Amin, *Accumulation on a World Scale.*
51. David F. Noble, *Progress Without People: New Technology, Unemployment, and the Message of Resistance* (Between The Lines, 1995), 65.
52. Emmanuel Arghiri, *Unequal Exchange: A Study of the Imperialism of Trade* (Monthly Review Press, 1972); Alf Hornborg, "Towards an Ecological Theory of Unequal Exchange: Articulating World System Theory and Ecological Economics," *Ecological Economics* 25, no. 1 (1998): 127–36.

CHAPTER 3

1. Ali Kadri, *Arab Development Denied: Dynamics of Accumulation by Wars of Encroachment* (Anthem Press, 2015); Sara Roy, *The Gaza Strip: The Political Economy of De-Development* (Institute for Palestine Studies, 2016); Utsa Patnaik and Prabhat Patnaik, *A Theory of Imperialism* (Columbia University Press, 2016); Max Ajl, "Does the Arab Region Have an Agrarian Question?," *Journal of Peasant Studies*, 2020, https://doi.org/10.1080/03066150.2020.1753706; Richard White, *The Roots of Dependency: Subsistence, Environment, and Social Change Among the Choctaws, Pawnees, and Navajos* (University of Nebraska Press, 1988).
2. Gallup Inc, "Preference for Environment Over Economy Largest Since 2000," Gallup.com, April 4, 2019, https://news.gallup.com/poll/248243/preference-environment-economy-largest-2000.aspx.
3. Max Ajl, "Degrowth Considered," *Brooklyn Rail*, September 2018. This is a statement more about theory than about implied alliances; the website *Uneven Earth*, for example, has published a lot on climate colonialism and collects and distributes a great deal of information about national-popular struggles as are ongoing in Venezuela.
4. Joan Martínez-Alier, *The Environmentalism of the Poor: A Study of Ecological Conflicts and Valuation* (Edward Elgar Publishing, 2003).
5. Giorgos Kallis, Christian Kerschner, and Joan Martinez-Alier, "The Economics of Degrowth," *Ecological Economics* 84 (December 2012): 172–80, https://doi.org/10.1016/j.ecolecon.2012.08.017; Christine Corlet Walker, "Review: *In Defense of Degrowth: Opinions and Manifestos*, by Giorgos Kallis, edited by Aaron Vansintjan, Uneven Earth Press, 2018," *Ecological Economics* 156 (2018): 431–2; Giacomo D'Alisa, Federico Demaria, and Giorgos Kallis, *Degrowth: A Vocabulary for a New Era* (Routledge, 2014).

6. "Socialism, Capitalism and the Transition Away from Fossil Fuels," *openDemocracy*, accessed January 30, 2020, www.opendemocracy.net/en/oureconomy/socialism-capitalism-and-transition-away-fossil-fuels/?fbclid=IwAR1tvCaorruCZ13b4IrrqpRc_V04VxcAX9aefmknPeTB7ucErZMMbn4koR8.
7. Keynyn Brysse et al., "Climate Change Prediction: Erring on the Side of Least Drama?," *Global Environmental Change* 23, no. 1 (February 1, 2013): 327–37, https://doi.org/10.1016/j.gloenvcha.2012.10.008.
8. Jasmine E. Livingston and Markku Rummukainen, "Taking Science by Surprise: The Knowledge Politics of the IPCC Special Report on 1.5 Degrees," *Environmental Science & Policy* 112 (October 1, 2020): 10–16, https://doi.org/10.1016/j.envsci.2020.05.020.
9. Anton Vaks et al., "Speleothems Reveal 500,000-Year History of Siberian Permafrost," *Science* 340, no. 6129 (2013): 183–86.
10. IPCC, "Chapter 2 – Global Warming of 1.5°C," 2018, www.ipcc.ch/sr15/chapter/chapter-2/.
11. J. Timmons Roberts and Bradley Parks, *A Climate of Injustice: Global Inequality, North-South Politics, and Climate Policy* (MIT Press, 2006), 77.
12. "Forced from Home: Climate-Fuelled Displacement," *Oxfam International*, December 2, 2019, https://www.oxfam.org/en/research/forced-home-climate-fuelled-displacement.
13. Rehad Desai, "Catastrophe Is upon Us–the Grim View from Southern Africa," *MR Online* (blog), February 13, 2020, https://mronline.org/2020/02/13/catastrophe-is-upon-us-the-grim-view-from-southern-africa/.
14. "Did Climate Change Cause the Flooding in the Midwest and Plains?," *Yale Climate Connections* (blog), April 2, 2019, www.yaleclimateconnections.org/2019/04/did-climate-change-cause-midwest-flooding/.
15. Natalie Delgadillo, "MAP: How Much Climate Change Will Cost Each U.S. County," accessed January 26, 2021, www.governing.com/archive/gov-counties-climate-change-damages-economic-effects-map.html.
16. "Chapter 2 – Global Warming of 1.5°C," 100, accessed February 18, 2020, www.ipcc.ch/sr15/chapter/chapter-2/.
17. Catriona McKinnon, "Runaway Climate Change: A Justice-Based Case for Precautions," *Journal of Social Philosophy* 40, no. 2 (2009): 187–203.
18. D. Kriebel et al., "The Precautionary Principle in Environmental Science," *Environmental Health Perspectives* 109, no. 9 (September 1, 2001): 871–76, https://doi.org/10.1289/ehp.01109871.
19. Robert Pollin, "De-Growth vs a Green New Deal," *New Left Review* II, no. 112 (2018): 5–25.
20. "A Response to Pollin and Chomsky: We Need a Green New Deal without Growth," Jason Hickel, accessed February 1, 2021, www.jasonhickel.org/blog/2020/10/19/we-need-a-green-new-deal-without-growth.
21. Ted Trainer, "Estimating the EROI of Whole Systems for 100% Renewable Electricity Supply Capable of Dealing with Intermittency," *Energy Policy* 119 (August 1, 2018): 648–53, https://doi.org/10.1016/j.enpol.2018.04.045.

22. Ted Trainer, "Can Australia Run on Renewable Energy? The Negative Case," *Energy Policy*, Special Section: Past and Prospective Energy Transitions – Insights from History, 50 (November 1, 2012): 306–14, https://doi.org/10.1016/j.enpol.2012.07.024.
23. Antoine Beylot et al., "Mineral Raw Material Requirements and Associated Climate-Change Impacts of the French Energy Transition by 2050," *Journal of Cleaner Production* 208 (January 2019): 1198–1205, https://doi.org/10.1016/j.jclepro.2018.10.154.
24. Ryan Stock and Trevor Birkenholtz, "The Sun and the Scythe: Energy Dispossessions and the Agrarian Question of Labor in Solar Parks," *Journal of Peasant Studies*, accessed November 24, 2020, www.tandfonline.com/doi/abs/10.1080/03066150.2019.1683002.
25. Owen Dowling, "The Political Economy of Super-Exploitation in Congolese Mineral Mining" (Cambridge University, 2020).
26. Eric Bonds and Liam Downey, "Green" Technology and Ecologically Unequal Exchange: The Environmental and Social Consequences of Ecological Modernization in the World-System," *Journal of World-Systems Research* 18, no. 2 (August 26, 2012): 167–86, https://doi.org/10.5195/jwsr.2012.482.
27. Roldan Muradian, Mauricio Folchi, and Joan Martinez-Alier, "'Remoteness' and Environmental Conflicts: Some Insights from the Political Ecology and Economic Geography of Copper," n.d., 20.
28. Max Ajl, "Stories About Oil and War," *Journal of Labor and Society*, 2021.
29. Tim Crownshaw, "Energy and the Green New Deal – Uneven Earth," accessed February 10, 2020, http://unevenearth.org/2020/01/energy-and-the-green-new-deal/.
30. Pollin, "De-Growth vs a Green New Deal."
31. Robert Pollin, "De-Growth vs a Green New Deal," *New Left Review*, no. 112 (August 2018), https://newleftreview.org/issues/ii112/articles/robert-pollin-de-growth-vs-a-green-new-deal.
32. Jennifer E. Givens, Xiaorui Huang, and Andrew K. Jorgenson, "Ecologically Unequal Exchange: A Theory of Global Environmental Injustice," *Sociology Compass* 13, no. 5 (2019): e12693.
33. Robert E. B. Lucas, David Wheeler, and Hemamale Hettige, *Economic Development, Environmental Regulation, and the International Migration of Toxic Industrial Pollution, 1960–88*, vol. 1062 (World Bank Publications, 1992).
34. UNCTAD, *Financing a Global Green New Deal*, Trade and Development Report (United Nations, 2019).
35. Stan Cox, *The Green New Deal and Beyond: Ending the Climate Emergency While We Still Can* (City Lights Books, 2020), 97–98.
36. "Energy Inequality – Conceptual Notes and Declarations – IIASA," accessed January 26, 2021, https://iiasa.ac.at/web/home/research/alg/energy-inequality.html.

37. Stan Cox, "Cornucopian Renewable-Energy Claims Leave Poor Nations in the Dark," *Resilience*, February 26, 2018, www.resilience.org/stories/2018-02-26/cornucopian-renewable-energy-claims-leave-poor-nations-dark/.
38. Joel Millward-Hopkins et al., "Providing Decent Living with Minimum Energy: A Global Scenario," *Global Environmental Change* 65 (November 1, 2020): 102168, https://doi.org/10.1016/j.gloenvcha.2020.102168.
39. Kris De Decker, "History and Future of the Compressed Air Economy," *LOW-TECH MAGAZINE*, accessed January 25, 2021, www.lowtechmagazine.com/2018/05/history-and-future-of-the-compressed-air-economy.html.
40. Working Group 13, "Final Conclusions Working Group 13: Intercultural Dialogue to Share Knowledge, Skills and Technologies," *World People's Conference on Climate Change and the Rights of Mother Earth* (blog), April 29, 2010, https://pwccc.wordpress.com/2010/04/29/final-conclusions-working-group13-intercultural-dialogue-to-share-knowledge-skills-and-technologies/.

CHAPTER 4

1. "Global Warming of 1.5°C," accessed January 27, 2021, www.ipcc.ch/sr15/.
2. Basil Davidson, *Scenes from Anti-Nazi War* (New York: Monthly Review Press, 1981).
3. Peter Gowan, *The Global Gamble : Washington's Faustian Bid for World Dominance* (London ; New York : Verso, 1999).
4. Patnaik and Patnaik, *A Theory of Imperialism*; Max Ajl, "The Arab Nation, The Chinese Model, and Theories of Self-Reliant Development," in *Non-Nationalist Forms of Nation-Based Radicalism: Nation beyond the State and Developmentalism*, ed. Ilker Corut and Joost Jongerden (Routledge, 2021) and sources cited.
5. Utsa Patnaik, *The Republic of Hunger and Other Essays* (Merlin Press, 2007).
6. John T. Callaghan, *The Retreat of Social Democracy* (Manchester University Press, 2000); Frances Fox Piven and Richard Cloward, *Poor People's Movements: Why They Succeed, How They Fail* (Knopf Doubleday Publishing Group, 2012).
7. IRENA, "Renewable Energy and Jobs – Annual Review 2019," https://www.irena.org/publications/2019/Jun/Renewable-Energy-and-Jobs-Annual-Review-2019.
8. Laura Pérez-Sánchez, Raúl Velasco-Fernández, and Mario Giampietro, "The International Division of Labor and Embodied Working Time in Trade for the US, the EU and China," *Ecological Economics* 180 (February 1, 2021): 106909, https://doi.org/10.1016/j.ecolecon.2020.106909.
9. Alexander Dunlap, *Renewing Destruction: Wind Energy Development, Conflict and Resistance in a Latin American Context* (Rowman & Littlefield International, 2019).
10. Leah Temper et al., "Movements Shaping Climate Futures: A Systematic Mapping of Protests against Fossil Fuel and Low-Carbon Energy Projects,"

Environmental Research Letters 15, no. 12 (November 2020): 123004, https://doi.org/10.1088/1748-9326/abc197.

11. Gilio-Whitaker, *As Long As Grass Grows*.
12. Kim Scipes, *AFL-CIO's Secret War Against Developing Country Workers: Solidarity Or Sabotage?* (Lexington Books, 2011).
13. David J. Hess, *Good Green Jobs in a Global Economy: Making and Keeping New Industries in the United States* (MIT Press, 2012), 53.
14. Green Party of the United States, "The Green New Deal," Green Party of the United States, January 21, 2019, https://gpus.org/organizing-tools/the-green-new-deal/.
15. Ocasio-Cortez, "Text – H.Res.109 – 116th Congress (2019–2020)."
16. Benjamin Selwyn, "A Green New Deal for Agriculture: For, within, or against Capitalism?," *The Journal of Peasant Studies* (January 29, 2021): 1–29, https://doi.org/10.1080/03066150.2020.1854740.
17. "AOC Says She Follows Democratic Leadership on Issue of US Intervention in Venezuela," accessed January 15, 2021, www.telesurenglish.net//news/AOC-Says-She-Follows-Democratic-Leadership-on-Issue-of-US-Intervention-in-Venezuela-20190504-0029.html.
18. "Pelosi Statement on the Situation in Venezuela," Speaker Nancy Pelosi, February 8, 2019, www.speaker.gov/newsroom/2819-2.
19. Mitropoulos, "Playing With Fire."
20. Climate Finance Leadership Initiative, "Financing the Low-Carbon Future: A Private-Sector View on Mobilizing Climate Finance."
21. "U.S. Will Pay into Climate Fund, but Not Reparations: Todd Stern," *Reuters*, December 9, 2009, www.reuters.com/article/us-climate-copenhagen-stern-idUSTRE5B82R220091209.
22. https://livingwage.mit.edu/articles/61-new-living-wage-data-for-now-available-on-the-tool#:~:text=The%20living%20wage%20in%20the,wage%20for%20most%20American%20families.
23. Cited in Meghashyam Mali, "Ocasio-Cortez: Democratic Socialism Is 'Part of What I Am, It's Not All of What I Am,'" Text, *TheHill*, July 1, 2018, https://thehill.com/homenews/sunday-talk-shows/395073-ocasio-cortez-socialism-is-part-of-what-i-am-its-not-all-of-what-i.
24. "Saikat Chakrabarti on Twitter," Twitter, accessed September 24, 2020, https://twitter.com/saikatc/status/1176624781686378501.
25. "AOC's Chief of Staff Admits the Green New Deal Is Not about Climate Change," accessed September 24, 2020, https://news.yahoo.com/aoc-chief-staff-admits-green-124408358.html.
26. Jesse Goldstein, *Planetary Improvement: Cleantech Entrepreneurship and the Contradictions of Green Capitalism* (MIT Press, 2018).
27. "AOC's Green New Deal Starts Strong," accessed January 19, 2021, https://jacobinmag.com/2019/02/aoc-green-new-deal-pelosi-democrats-climate.
28. Oscar Reyes, *Change Finance, Not the Climate* (Transnational Institute, 2020), 33–34, 90, www.tni.org/en/changefinance.

29. Kate Aronoff et al., *A Planet to Win: Why We Need a Green New Deal* (Verso Books, 2019); Naomi Klein, *On Fire: The (Burning) Case for a Green New Deal* (Simon & Schuster, 2019).
30. "The Green New Deal," Bernie Sanders Official Website, accessed February 1, 2021, https://berniesanders.com/issues/green-new-deal/.
31. Noam Chomsky and Robert Pollin, *Climate Crisis and the Global Green New Deal: The Political Economy of Saving the Planet* (Verso Books, 2020).
32. Mitropoulos, "Playing With Fire."
33. "350.Org," Rockefeller Brothers Fund, August 25, 2015, www.rbf.org/grantees/350org; Cécile, "Why ECF - European Climate Foundation," https://europeanclimate.org/, accessed September 26, 2020, https://europeanclimate.org/why-ecf/; "2019 Annual Report: Financial Data," 350.org, accessed September 26, 2020, https://350.org/2019-annual-report-financials/.
34. "The Arkay Foundation," accessed September 26, 2020, www.arkayfoundation.org/reducing_2017.html.
35. Bill McKibben, "But There Now Seems the Real Possibility of Concerted Action across the Federal Government to Make Sweeping Change – If, of Course, There Are Movements Prodding, and Opening up Space, and Cheering Accomplishments to Build Momentum. That's the Work of the Rest of Us.," Tweet, @billmckibben (blog), December 17, 2020, https://twitter.com/billmckibben/status/1339689008646004744.
36. "Plan for Climate Change and Environmental Justice | Joe Biden," Joe Biden for President: Official Campaign Website, accessed January 26, 2021, https://joebiden.com/climate-plan/.
37. "Extinction Rebellion UK on Twitter," Twitter, accessed September 26, 2020, https://twitter.com/XRebellionUK/status/1300794775138906114.
38. Naomi Klein, *On Fire: The (Burning) Case for a Green New Deal* (Simon & Schuster, 2019).
39. Benjamin Selwyn, "A Green New Deal for Agriculture: For, within, or against Capitalism?," *The Journal of Peasant Studies* (January 29, 2021): 1–29, https://doi.org/10.1080/03066150.2020.1854740.
40. Cira Pascual Marquina and Chris Gilbert, *Venezuela, the Present as Struggle: Voices from the Bolivarian Revolution* (Monthly Review Press, 2020).
41. Kate Aronoff et al., *A Planet to Win: Why We Need a Green New Deal* (Verso Books, 2019).
42. LVC, "La Via Campesina in Action for Climate Justice," FAO, 2019, www.fao.org/agroecology/database/detail/en/c/1199383/.
43. "The Covid-19 Pandemic Shows We Must Transform the Global Food System | Jan Dutkiewicz, Astra Taylor and Troy Vettese," *The Guardian*, April 16, 2020, www.theguardian.com/commentisfree/2020/apr/16/coronavirus-covid-19-pandemic-food-animals.
44. Kate Aronoff et al., *A Planet to Win: Why We Need a Green New Deal* (Verso Books, 2019).

45. Keston Perry, "Financing a Global Green New Deal: Between Techno-Optimist Renewable Energy Futures and Taming Financialization for a New 'Civilizing' Multilateralism," *Development and Change*, Forthcoming, 5.

CHAPTER 5

1. Ebenezer Howard, *Garden Cities of To-Morrow* (Routledge, 2013); Ivan Illich, *Tools for Conviviality* (Boyars, 1985); Ismail-Sabri Abdallah, "Dépaysanisation Ou Développement Rural? Un Choix Lourd de Conséquences," in *IFDA Dossier*, vol. 9 (Nyon, Switzerland: International Foundation for Development Alternatives, 1979), 1–15; Lewis Mumford, "Authoritarian and Democratic Technics," *Technology and Culture* 5, no. 1 (1964): 1–8; Ursula K. Le Guin, *The Dispossessed: An Ambiguous Utopia* (EOS, 1999).
2. Samir Amin, *Delinking: Towards a Polycentric World* (Zed Books, 1990); Samir Amin, *The Future of Maoism* (Monthly Review Pr, 1983); Fawzy Mansour, "Third World Revolt and Self-Reliant Auto-Centered Strategy of Development," in *Toward a New Strategy for Development: A Rothko Chapel Colloquium* (Pergamon, 1979); Mohamed Dowidar, "The Self-Reliance Strategy of Development and the New International Economic Order," *Mondes En Développement* 26 (1979): 249–54; Abdul Rahman Mohamed Babu, *The Future That Works: Selected Writings of A.M. Babu* (Africa World Press, 2002), 18–23; Max Ajl, "Auto-Centered Development and Indigenous Technics: Slaheddine El-Amami and Tunisian Delinking," *Journal of Peasant Studies* 46, no. 6 (2019): 1240–63; Max Ajl, "Delinking, Food Sovereignty, and Populist Agronomy: Notes on an Intellectual History of the Peasant Path in the Global South," *Review of African Political Economy* 45, no. 155 (2018): 64–84.
3. Kali Akuno, "Build and Fight: The Program and Strategy of Cooperation Jackson," in *Jackson Rising: The Struggle for Economic Democracy and Black Self-Determination in Jackson, Mississippi*, ed. Kali Akuno and Ajamu Nangwaya (Daraja Press, 2017), 3–41.
4. Adel Samara, *Al-Tanmīyya b-al-Ḥamāyya Al-Sha'bīyya* (Al-Quds: Markaz al-Zahrā' lil-Dirāsāt w al-Ābḥāth, 1990); Adel Samara, *Beyond De-Linking: Development by. Popular Protection vs. Development by State* (Palestine Research and Publishing Foundation, 2005); Max Ajl, "Development by Popular Protection and Tunisia: The Case of Tataouine," *Globalizations* 16, no. 7 (2019): 1215–31.
5. Ivan Illich, *Tools for Conviviality* (Harper & Row, 1973).
6. Jan Douwe van der Ploeg, *Born from Within: Practice and Perspectives of Endogenous Rural Development* (Uitgeverij Van Gorcum, 1994).
7. Stefania Barca, "The Labor(s) of Degrowth," *Capitalism Nature Socialism* 30, no. 2 (April 3, 2019): 207–16, https://doi.org/10.1080/10455752.2017.1373300.
8. Leonidas Oikonomakis, "Vio.Me: The Greek Factory without Bosses — an Interview," *ROAR Magazine*, accessed November 24, 2020, https://roarmag.org/essays/vio-me-factory-without-bosses/.

9. Frederick H. Buttel, "The Treadmill of Production: An Appreciation, Assessment, and Agenda for Research," *Organization & Environment* 17, no. 3 (September 1, 2004): 323–36, https://doi.org/10.1177/1086026604267938.
10. Anders Hayden and John M. Shandra, "Hours of Work and the Ecological Footprint of Nations: An Exploratory Analysis," *Local Environment* 14, no. 6 (July 1, 2009): 575–600, https://doi.org/10.1080/13549830902904185.
11. Jeanette W. Chung and David O. Meltzer, "Estimate of the Carbon Footprint of the US Health Care Sector," *JAMA* 302, no. 18 (November 11, 2009): 1970, https://doi.org/10.1001/jama.2009.1610.
12. Carol Lynn Esposito et al., "Against All Odds: Cuba Achieves Healthcare for All – an Analysis of Cuban Healthcare," *JNY State Nurses Assoc* 45, no. 1 (2016): 29–38.
13. Salimah Valiani, *Rethinking Unequal Exchange: The Global Integration of Nursing Labour Markets* (University of Toronto Press, 2012).
14. Jose Maria Sison, *Foundation for Resuming the Philippine Revolution: Selected Writings, 1968 to 1972* (International Network for Philippine Studies, 2013).
15. World Bank, selected indicators.
16. Selma James, *Sex, Race and Class, the Perspective of Winning: A Selection of Writings 1952–2011* (PM Press, 2012); Silvia Federici, *Revolution at Point Zero: Housework, Reproduction, and Feminist Struggle* (PM Press, 2012).
17. Alf Hornborg, "Zero-Sum World: Challenges in Conceptualizing Environmental Load Displacement and Ecologically Unequal Exchange in the World-System," *International Journal of Comparative Sociology* 50, no. 3–4 (2009): 237–62.
18. Sandra Halperin, *Re-Envisioning Global Development: A Horizontal Perspective* (Routledge, 2013); Janet L. Abu-Lughod, *Before European Hegemony: The World System A.d. 1250–1350* (Oxford University Press, 1991); Neil Brenner, *New State Spaces: Urban Governance and the Rescaling of Statehood* (Oxford University Press, 2004).
19. Ricardo Jacobs, "An Urban Proletariat with Peasant Characteristics: Land Occupations and Livestock Raising in South Africa," *The Journal of Peasant Studies* 45, no. 5–6 (2018): 884–903.
20. "One Way To Close The Black Homeownership Gap: Housing As Reparations," KQED, accessed November 21, 2020, www.kqed.org/news/11841801/what-we-owe-housing-as-reparations.
21. Nandini Bagchee and The Advanced Design Students (Spring 2019) CUNY, "Building a Transition City_ Landscape Online Version | Jackson | Sustainability" (CUNY), accessed November 24, 2020, www.scribd.com/document/431893329/Building-a-Transition-City-Landscape-Online-Version.
22. Max Ajl, "The Hypertrophic City versus the Planet of Fields," in *Implosions/Explosions. Berlin: Jovis*, ed. Neil Brenner (Jovis, 2014), 533–50; Ivan Kremnev, "The Journey of My Brother Alexei to the Land of Peasant Utopia," *Journal of Peasant Studies* 4, no. 1 (1976): 63–108, https://doi.org/10.1080/03066157608438004.

23. Rowan Moore, "Wasteful, Damaging and Outmoded: Is it Time to Stop Building Skyscrapers?," *The Guardian*, July 11, 2020, www.theguardian.com/artanddesign/2020/jul/11/skyscrapers-wasteful-damaging-outmoded-time-to-stop-tall-buildings.
24. Seungtaek Lee and Wai Oswald Chong, "Causal Relationships of Energy Consumption, Price, and CO2 Emissions in the US Building Sector," *Resources, Conservation and Recycling* 107 (2016): 220–26.
25. J. F. Correal, "State-of-the-Art of Practice in Colombia on Engineered Guadua Bamboo Structures," in *Modern Engineered Bamboo Structures: Proceedings of the Third International Conference on Modern Bamboo Structures (ICBS 2018), June 25–27, 2018, Beijing, China* (CRC Press, 2019), 23.
26. "Feature// ShamsArd – Furniture," *HKZ|MENA Design Magazine* (blog), September 3, 2013, www.herskhazeen.com/feature-shamsard-furniture/.
27. Zach Mortice, "Bamboo Transcends the Tropics for Carbon-Negative Construction," *Redshift EN* (blog), August 7, 2019, www.autodesk.com/redshift/bamboo-construction/.
28. David Roberts, "The Hottest New Thing in Sustainable Building Is, Uh, Wood," *Vox*, January 15, 2020, www.vox.com/energy-and-environment/2020/1/15/21058051/climate-change-building-materials-mass-timber-cross-laminated-clt.
29. Tarun Jami, Deepak M. E Phd, and Yadvendra Agrawal, "Hemp Concrete: Carbon Negative Construction," *Emerging Materials Research* 5 (July 1, 2016), https://doi.org/10.1680/jemmr.16.00122; Tuomas Mattila et al., "Is Biochar or Straw-Bale Construction a Better Carbon Storage from a Life Cycle Perspective?," *Process Safety and Environmental Protection*, Special Issue: Negative emissions technology, 90, no. 6 (November 1, 2012): 452–58, https://doi.org/10.1016/j.psep.2012.10.006.
30. OAR US EPA, "Sources of Greenhouse Gas Emissions," Overviews and Factsheets, US EPA, December 29, 2015, www.epa.gov/ghgemissions/sources-greenhouse-gas-emissions.
31. "Transport – IPCC," accessed January 16, 2021, www.ipcc.ch/report/ar5/wg3/transport/.
32. "Leading Scientists Set out Resource Challenge of Meeting Net Zero Emissions in the UK by 2050," accessed January 16, 2021, www.nhm.ac.uk/press-office/press-releases/leading-scientists-set-out-resource-challenge-of-meeting-net-zer.html.
33. CBS News, "Apple, Google, Microsoft, Tesla and Dell Sued over Child-Mined Cobalt from Africa," December 17, 2019, www.cbsnews.com/news/apple-google-microsoft-tesla-dell-sued-over-cobalt-mining-children-in-congo-for-batteries-2019-12-17/.
34. Ivan Illich, *Tools for Conviviality* (Boyars, 1985).
35. Bureau of Transportation Statistics, "Freight Facts and Figures 2017" (Department of Transportation, 2017).
36. Matthew Beedham, "Swedes to Build Wind-Powered Transatlantic Cargo Ship (Yes, It's a Sailboat)," Shift | *The Next Web*, September 10, 2020,

https://thenextweb.com/shift/2020/09/10/swedes-boat-powered-by-wind-sailboat-ship-cargo-transatlantic/.

37. Paul A. T. Higgins and Millicent Higgins, "A Healthy Reduction in Oil Consumption and Carbon Emissions," *Energy Policy* 33, no. 1 (January 1, 2005): 1–4, https://doi.org/10.1016/S0301-4215(03)00201-5.
38. Colin A. M. Duncan, "On Identifying a Sound Environmental Ethic in History: Prolegomena to Any Future Environmental History," *Environmental History Review* (1991), 21.
39. James Salazar and Jamie Meil, "Prospects for Carbon-Neutral Housing: The Influence of Greater Wood Use on the Carbon Footprint of a Single-Family Residence," *Journal of Cleaner Production* 17, no. 17 (November 1, 2009): 1563–71, https://doi.org/10.1016/j.jclepro.2009.06.006.
40. Kris De Decker, *Low-Tech Magazine 2012–2018* (Kris De Decker, 2019).

CHAPTER 6

1. Ajl, "The Hypertrophic City versus the Planet of Fields." Special thanks to Eric Holt-Giménez for suggesting I rework this chapter.
2. Kyle Powys Whyte, "Indigenous Climate Justice and Food Sovereignty," in *Indigenous Food Sovereignty in the United States: Restoring Cultural Knowledge, Protecting Environments, and Regaining Health*, ed. Devon A. Mihesuah and Elizabeth Hoover (University of Oklahoma Press, 2019), 327.
3. Colin A. M. Duncan, *Centrality of Agriculture: Between Humankind and the Rest of Nature* (McGill-Queen's Press-MQUP, 1996), 181.
4. Raj Patel, "Food Sovereignty," *The Journal of Peasant Studies* 36, no. 3 (2009): 663–706.
5. William M. Denevan, "The Pristine Myth: The Landscape of the Americas in 1492," *Annals of the Association of American Geographers* 82, no. 3 (1992): 369–85.
6. Kat Anderson, *Tending the Wild: Native American Knowledge and the Management of California's Natural Resources* (University of California Press, 2005), 245.
7. Dennis Michael Warren, "Indigenous Agricultural Knowledge, Technology, and Social Change," in *Sustainable Agriculture in the American Midwest: Lessons from the Past, Prospects for the Future*, ed. Gregory McIsaac and William R. Edwards (University of Illinois Press, 1994), 35–53.
8. Jane Mt Pleasant, "Food Yields and Nutrient Analyses of the Three Sisters: A Haudenosaunee Cropping System," *Ethnobiology Letters* 7, no. 1 (2016): 87–98.
9. Natalie Kurashima, Lucas Fortini, and Tamara Ticktin, "The Potential of Indigenous Agricultural Food Production under Climate Change in Hawai'i," *Nature Sustainability* 2, no. 3 (2019): 191–99. Thanks to Albie Miles for discussion on this.
10. Gilio-Whitaker, *As Long As Grass Grows*, 76.

11. Harriet Friedmann and Philip McMichael, "Agriculture and the State System," *Landwirtschaft Und Staatliches System: Aufstieg Und Niedergang Der Nationalen Landwirtschaft von 1870 Biz Zur Gegenwart.* 29, no. 2 (April 1989): 93.
12. "Research Findings from the Land Inequality Initiative," global, accessed February 8, 2021, www.landcoalition.org/en/uneven-ground/.
13. Benjamin E. Graeub et al., "The State of Family Farms in the World," *World Development* 87 (November 1, 2016): 1–15, https://doi.org/10.1016/j.worlddev.2015.05.012.
14. Miguel A. Altieri and Victor Manuel Toledo, "The Agroecological Revolution in Latin America: Rescuing Nature, Ensuring Food Sovereignty and Empowering Peasants," *Journal of Peasant Studies* 38, no. 3 (2011): 587–612, www.tandfonline.com/doi/abs/10.1080/03066150.2011.582947; Ashlesha Khadse et al., "Taking Agroecology to Scale: The Zero Budget Natural Farming Peasant Movement in Karnataka, India," *The Journal of Peasant Studies* 45, no. 1 (January 2, 2018): 192–219, https://doi.org/10.1080/03066150.2016.1276450.
15. Leonard Dudley and Roger J. Sandilands, "The Side Effects of Foreign Aid: The Case of Public Law 480 Wheat in Colombia," *Economic Development and Cultural Change* 23, no. 2 (1975): 325–36; Max Ajl, "Farmers, Fellaga, and Frenchmen" (PhD, Cornell University, 2019); Harriet Friedmann et al., "The Origins of Third World Food Dependence.," *Food Question: Profits versus People?* (1990), 13–31.
16. Arindam Banerjee, "The Longer 'Food Crisis' and Consequences for Economic Theory and Policy in the South," in *Rethinking the Social Sciences with Sam Moyo*, ed. Praveen Jha, Paris Yeros, and Walter Chambati (New Delhi: Tulika Books, 2020), 152–79.
17. Eric Holt-Giménez, *A Foodie's Guide to Capitalism* (Monthly Review Press, 2017).
18. Utsa Patnaik, "Revisiting the 'Drain', or Transfers from India to Britain in the Context of Global Diffusion of Capitalism," in *Agrarian and Other Histories: Essays for Binay Bhushan Chaudhuri*, ed. Shubhra Chakrabarti and Utsa Patnaik (Tulika Books, 2017), 277–318; Alec Gordon, "Netherlands East Indies: The Large Colonial Surplus of Indonesia, 1878–1939," *Journal of Contemporary Asia* 40, no. 3 (August 1, 2010): 425–43, https://doi.org/10.1080/00472331003798392; Alec Gordon, "A Last Word: Amendments and Corrections to Indonesia's Colonial Surplus 1880–1939," *Journal of Contemporary Asia* 48, no. 3 (May 27, 2018): 508–18, https://doi.org/10.1080/00472336.2018.1433865; Utsa Patnaik, "Profit Inflation, Keynes and the Holocaust in Bengal, 1943–44," *Economic & Political Weekly* 53, no. 42 (2018): 33.
19. Utsa Patnaik, *The Republic of Hunger and Other Essays* (Merlin Press, 2007); Christian Zlolniski, *Made in Baja* (University of California Press, 2019), www.ucpress.edu/book/9780520300637/made-in-baja.

20. Patnaik and Patnaik, *A Theory of Imperialism*; Utsa Patnaik, "The Origins and Continuation of First World Import Dependence on Developing Countries for Agricultural Products," *Agrarian South: Journal of Political Economy* 4, no. 1 (April 1, 2015): 1–21; Sam Moyo and Paris Yeros, "Intervention: The Zimbabwe Question and the Two Lefts," *Historical Materialism* 15 (August 31, 2007): 171–204; Freedom Mazwi and George T. Mudimu, "Why are Zimbabwe's Land Reforms Being Reversed?," *Economic and Political Weekly* 54 (August 13, 2019); Ana Felicien, Christina Schiavoni, and Liccia Romero, "The Politics of Food in Venezuela," *Monthly Review*, June 2018, https://monthlyreview.org/2018/06/01/the-politics-of-food-in-venezuela/.
21. Caspar A. Hallmann et al., "More than 75 Percent Decline over 27 Years in Total Flying Insect Biomass in Protected Areas," *PLOS ONE* 12, no. 10 (October 18, 2017): e0185809, https://doi.org/10.1371/journal.pone.0185809; Annie Shattuck, "Toxic Uncertainties and Epistemic Emergence: Understanding Pesticides and Health in Lao PDR," *Annals of the American Association of Geographers* (June 10, 2020): 1–15, https://doi.org/10.1080/24694452.2020.1761285; Divya Sharma, "Techno-Politics, Agrarian Work and Resistance in Post-Green Revolution Punjab, India" (Dissertation, Cornell University, 2017); Wolfgang Boedeker et al., "The Global Distribution of Acute Unintentional Pesticide Poisoning: Estimations Based on a Systematic Review," *BMC Public Health* 20, no. 1 (December 7, 2020): 1875, https://doi.org/10.1186/s12889-020-09939-0.
22. Cited in Joan Martinez-Alier, "The EROI of Agriculture and Its Use by the Via Campesina," *Journal of Peasant Studies* 38, no. 1 (2011): 145–60, https://doi.org/10.1080/03066150.2010.538582.
23. Gerald Leach, "Energy and Food Production," *Food Policy* 1, no. 1 (1975): 62–73.
24. Mads V. Markussen and Hanne Østerg\aard, "Energy Analysis of the Danish Food Production System: Food-EROI and Fossil Fuel Dependency," *Energies* 6, no. 8 (2013): 4177, 4179.
25. Nathan Pelletier et al., "Energy Intensity of Agriculture and Food Systems," *Annual Review of Environment and Resources* 36, no. 1 (2011): 223–46, https://doi.org/10.1146/annurev-environ-081710-161014.
26. Eric Holt-Giménez, *A Foodie's Guide to Capitalism* (Monthly Review Press, 2017), 195–200.
27. Wendell Berry, *The Unsettling of America: Culture and Agriculture* (San Val, Incorporated, 1996); Linda Lobao and Katherine Meyer, "The Great Agricultural Transition: Crisis, Change, and Social Consequences of Twentieth Century US Farming," *Annual Review of Sociology* 27, no. 1 (2001): 108–9, https://doi.org/10.1146/annurev.soc.27.1.103.
28. Trish Hernandez and Susan Gabbard, "Findings from the National Agricultural Workers Survey (NAWS) 2015-2016," NAWS Research Report, accessed February 7, 2021, www.dol.gov/sites/dolgov/files/ETA/naws/pdfs/NAWS_Research_Report_13.pdf; Ronald L. Mize and Alicia C. S. Swords,

Consuming Mexican Labor: From the Bracero Program to NAFTA (University of Toronto Press, 2010).

29. Carrie Freshour, "'Ain't No Life for a Mother': Racial Capitalism and the Making of Poultry Workers in the US South" (Dissertation, Cornell University, 2018).
30. Nick Cullather, *The Hungry World* (Harvard University Press, 2010).
31. "The State of Food Security and Nutrition in the World 2020 | FAO | Food and Agriculture Organization of the United Nations," 169, accessed February 7, 2021, https://doi.org/10.4060/CA9692EN.
32. "Yemen Emergency | World Food Programme," accessed February 7, 2021, www.wfp.org/emergencies/yemen-emergency.
33. "The State of Food Security and Nutrition in the World 2020 | FAO | Food and Agriculture Organization of the United Nations," 67, accessed February 7, 2021, https://doi.org/10.4060/CA9692EN.
34. I discuss this here: Max Ajl, "How Much Will the US Way of Life © Have to Change? On the Future of Farming, Socialist Science, and Utopia," *Uneven Earth*, June 10, 2019, www.researchgate.net/publication/334508917_How_much_will_the_US_Way_of_Life_C_have_to_change_On_the_future_of_farming_socialist_science_and_utopia.
35. Eric Holt-Giménez, *A Foodie's Guide to Capitalism* (Monthly Review Press, 2017).
36. Via Campesina, "VIIth International Conference, Euskal Herria Declaration," *Via Campesina English* (blog), July 26, 2017, https://viacampesina.org/en/viith-international-conference-la-via-campesina-euskal-herria-declaration/.
37. John Bellamy Foster, "The Meaning of Work in a Sustainable Society," *Monthly Review*, September 2017.
38. Duncan, *Centrality of Agriculture*, 181–82.
39. Ivette Perfecto, John H. Vandermeer, and Angus Lindsay Wright, *Nature's Matrix: Linking Agriculture, Conservation and Food Sovereignty* (Earthscan, 2009).
40. Giovanni Tamburini et al., "Agricultural Diversification Promotes Multiple Ecosystem Services without Compromising Yield," *Science Advances* 6, no. 45 (2020): eaba1715.
41. The following discussion draws from Peter Rosset and Miguel A. Altieri, *Agroecology: Science and Politics* (Practical Action Publishing, 2017).
42. Habib Ayeb, *Gabes Labess [All Is Well in Gabes]* (Tunis, Tunisia, 2014), www.youtube.com/watch?v=j_wkggqYCBg.
43. Miguel A. Altieri and Clara I. Nicholls, "The Adaptation and Mitigation Potential of Traditional Agriculture in a Changing Climate," *Climatic Change* 140, no. 1 (2017): 33–45; Eric Holt-Giménez, "Measuring Farmers' Agroecological Resistance after Hurricane Mitch in Nicaragua: A Case Study in Participatory, Sustainable Land Management Impact Monitoring," *Agriculture, Ecosystems & Environment* 93, no. 1–3 (December 2002): 87–105, https://doi.org/10.1016/S0167-8809(02)00006-3.

44. Rob Wallace, *Dead Epidemiologists: On the Origins of COVID-19* (Monthly Review Press, 2020), https://monthlyreview.org/product/dead-epidemiologists-on-the-origins-of-covid-19/.
45. Kayo Tajima, "The Marketing of Urban Human Waste in the Early Modern Edo/Tokyo Metropolitan Area," *Environnement Urbain / Urban Environment*, no. Volume 1 (September 9, 2007), http://journals.openedition.org/eue/1039; "Recycling Animal and Human Dung Is the Key to Sustainable Farming," *LOW-TECH MAGAZINE*, accessed October 21, 2020, www.lowtechmagazine.com/2010/09/recycling-animal-and-human-dung-is-the-key-to-sustainable-farming.html.
46. Ivette Perfecto, John H. Vandermeer, and Angus Lindsay Wright, *Nature's Matrix: Linking Agriculture, Conservation and Food Sovereignty* (Earthscan, 2009).
47. Christina Ergas, "Cuban Urban Agriculture as a Strategy for Food Sovereignty," *Monthly Review* 64, no. 10 (March 1, 2013), https://monthlyreview.org/2013/03/01/cuban-urban-agriculture-as-a-strategy-for-food-sovereignty/.
48. Nolwazi Zanele Khumalo, "An Assessment of the Contribution of Peri-Urban Agriculture on Household Food Security in Tongaat, EThekwini Municipality" (Thesis, University of Zululand, 2018), http://uzspace.unizulu.ac.za/xmlui/handle/10530/1817; Godwin Arku et al., "Africa's Quest for Food Security: What is the Role of Urban Agriculture?," 2012.
49. Mithun Saha and Matthew J. Eckelman, "Growing Fresh Fruits and Vegetables in an Urban Landscape: A Geospatial Assessment of Ground Level and Rooftop Urban Agriculture Potential in Boston, USA," *Landscape and Urban Planning* 165 (September 1, 2017): 130–41, https://doi.org/10.1016/j.landurbplan.2017.04.015; Francesco Orsini et al., "Exploring the Production Capacity of Rooftop Gardens (RTGs) in Urban Agriculture: The Potential Impact on Food and Nutrition Security, Biodiversity and Other Ecosystem Services in the City of Bologna," *Food Security* 6, no. 6 (December 1, 2014): 781–92, https://doi.org/10.1007/s12571-014-0389-6.
50. Aaron Vastinjan, "Cool People's Movements: Why Air Conditioners Aren't Good Enough for the Working Class," *The Ecologist*, September 24, 2018, https://theecologist.org/2018/sep/24/cool-peoples-movements-why-air-conditioners-arent-good-enough-working-class.
51. Muhammad Shafique, Xiaolong Xue, and Xiaowei Luo, "An Overview of Carbon Sequestration of Green Roofs in Urban Areas," *Urban Forestry & Urban Greening* 47 (January 1, 2020): 126515, https://doi.org/10.1016/j.ufug.2019.126515.
52. Flavie Mayrand and Philippe Clergeau, "Green Roofs and Green Walls for Biodiversity Conservation: A Contribution to Urban Connectivity?," *Sustainability* 10, no. 4 (April 2018): 985, https://doi.org/10.3390/su10040985.
53. James Wei Wang et al., "Building Biodiversity: Drivers of Bird and Butterfly Diversity on Tropical Urban Roof Gardens," *Ecosphere* 8, no. 9 (2017): e01905, https://doi.org/10.1002/ecs2.1905.

54. Gensuo Jia et al., "Land-Climate Interactions," in *Climate Change and Land*, IPCC, 2019, 189.
55. Bronson W. Griscom et al., "Natural Climate Solutions," *Proceedings of the National Academy of Sciences* 114, no. 44 (October 31, 2017): 11645–50, https://doi.org/10.1073/pnas.1710465114.
56. Hannah Ritchie and Max Roser, "CO_2 and Greenhouse Gas Emissions," *Our World in Data*, May 11, 2017, https://ourworldindata.org/co2-and-other-greenhouse-gas-emissions.
57. Keith Paustian et al., "Soil C Sequestration as a Biological Negative Emission Strategy," *Frontiers in Climate* 1 (2019), https://doi.org/10.3389/fclim.2019.00008.
58. Robert Blakemore, "Humic Carbon to Fix Food, Climate and Health," 2019.
59. Jamie Lorimer and Clemens Driessen, "Wild Experiments at the Oostvaardersplassen: Rethinking Environmentalism in the Anthropocene," *Transactions of the Institute of British Geographers* 39, no. 2 (2014): 169–81; Vera, *Grazing Ecology and Forest History*.
60. "The Greatest Climate-Protecting Technology Ever Devised," *Wired*, accessed February 7, 2021, www.wired.com/story/trees-plants-nature-best-carbon-capture-technology-ever/.
61. Bram Büscher and Robert Fletcher, "Towards Convivial Conservation," *Conservation & Society* 17, no. 3 (2019): 283–96; Bram Büscher et al., "Half-Earth or Whole Earth? Radical Ideas for Conservation, and Their Implications," *Oryx* 51, no. 3 (2017): 407–10.
62. Franciscus Wilhelmus Maria Vera, *Grazing Ecology and Forest History* (CABI publishing, 2000).
63. David E. Gilbert, "Laborers Becoming 'Peasants': Agroecological Politics in a Sumatran Plantation Zone," *The Journal of Peasant Studies* (July 15, 2019): 1–22, https://doi.org/10.1080/03066150.2019.1602521.
64. B. Schulz, B. Becker, and E. Götsch, "Indigenous Knowledge in a 'Modern' Sustainable Agroforestry System – a Case Study from Eastern Brazil," *Agroforestry Systems* 25, no. 1 (January 1, 1994): 59–69, https://doi.org/10.1007/BF00705706.
65. Budiadi and H. T. Ishii, "Comparison of Carbon Sequestration Between Multiple-Crop, Single-Crop and Monoculture Agroforestry Systems of Melaleuca in Java, Indonesia," *Journal of Tropical Forest Science* 22, no. 4 (2010): 378–88.
66. James M. Roshetko et al., "Carbon Stocks in Indonesian Homegarden Systems: Can Smallholder Systems Be Targeted for Increased Carbon Storage?," *American Journal of Alternative Agriculture* 17, no. 3 (September 2002): 138–48, https://doi.org/10.1079/AJAA200116.
67. Emanuela F. Gama-Rodrigues et al., "Carbon Storage in Soil Size Fractions Under Two Cacao Agroforestry Systems in Bahia, Brazil," *Environmental Management* 45, no. 2 (February 1, 2010): 274–83, https://doi.org/10.1007/s00267-009-9420-7.

68. K. Varsha et al., "High Density Silvopasture Systems for Quality Forage Production and Carbon Sequestration in Humid Tropics of Southern India," *Agroforestry Systems* (January 3, 2017), https://doi.org/10.1007/s10457-016-0059-0.
69. H. A. J. Gunathilake, "Coconut Based Farming Systems for Increasing Productivity and Profitability of Smallholder Coconut Plantation in Sri Lanka," n.d., https://library.apccsec.org/paneladmin/doc/20180406081240Dr.%20H.A.J.%20Gunathilake.91.pdf.
70. Ying Liu, "Evaluation of Breadfruit (Artocarpus Altilis and A. Altilis X A. Mariannensis) as a Dietary Protein Source" (University of British Columbia, 2016), https://doi.org/10.14288/1.0300337.
71. Ranjith P. Udawatta and Shibu Jose, "Agroforestry Strategies to Sequester Carbon in Temperate North America," *Agroforestry Systems* 86, no. 2 (October 2012): 225–42, https://doi.org/10.1007/s10457-012-9561-1.
72. H. E. Garrett et al., "Hardwood Silvopasture Management in North America," *Agroforestry Systems* 61–62, no. 1–3 (July 2004): 21–33, https://doi.org/10.1023/B:AGFO.0000028987.09206.6b.
73. T. G. Papachristou, P. D. Platis, and A. S. Nastis, "Foraging Behaviour of Cattle and Goats in Oak Forest Stands of Varying Coppicing Age in Northern Greece," *Small Ruminant Research*, Special Issue: Methodology nutrition and products quality in grazing sheep and goats, 59, no. 2 (August 1, 2005): 181–89, https://doi.org/10.1016/j.smallrumres.2005.05.006.
74. Simon Fairlie, *Meat: A Benign Extravagance* (Chelsea Green Publishing Company, 2010), 236–38.
75. Udawatta and Jose, "Agroforestry Strategies to Sequester Carbon in Temperate North America."
76. "Selected Writings from Members of WSU Breadlab | WSU Breadlab | Washington State University," accessed February 8, 2021, http://thebreadlab.wsu.edu/writings-from-the-bread-lab/. Thanks to Maywa Montenegro for the reference.
77. Tomek de Ponti, Bert Rijk, and Martin K. van Ittersum, "The Crop Yield Gap between Organic and Conventional Agriculture," *Agricultural Systems* 108 (April 1, 2012): 1–9, https://doi.org/10.1016/j.agsy.2011.12.004.
78. Wes Jackson, *New Roots for Agriculture* (University of Nebraska Press, 1980).
79. George Monbiot, "The Best Way to Save the Planet? Drop Meat and Dairy | George Monbiot," *The Guardian*, June 8, 2018, sec. Opinion, www.theguardian.com/commentisfree/2018/jun/08/save-planet-meat-dairy-livestock-food-free-range-steak; "Troy Vettese, To Freeze the Thames, NLR 111, May–June 2018," *New Left Review*, accessed February 11, 2020, https://newleftreview.org/issues/II111/articles/troy-vettese-to-freeze-the-thames; Troy Vettese, "The Last Man to Know Everything," Text, *Boston Review*, September 25, 2018.
80. I do not take up non-environmentally-related ethical arguments for veganism here.

81. Steven L. Dowhower et al., "Soil Greenhouse Gas Emissions as Impacted by Soil Moisture and Temperature under Continuous and Holistic Planned Grazing in Native Tallgrass Prairie," *Agriculture, Ecosystems & Environment* 287 (January 1, 2020): 106647, https://doi.org/10.1016/j.agee.2019.106647.
82. Pablo Manzano and Shannon White, "Intensifying Pastoralism May Not Reduce Greenhouse Gas Emissions: Wildlife-Dominated Landscape Scenarios as a Baseline in Life-Cycle Analysis," *Climate Research* 77 (February 21, 2019): 91–97, https://doi.org/10.3354/cr01555.
83. A. N. Hristov, "Historic, Pre-European Settlement, and Present-Day Contribution of Wild Ruminants to Enteric Methane Emissions in the United States," *Journal of Animal Science* 90, no. 4 (April 1, 2012): 1371–75, https://doi.org/10.2527/jas.2011-4539.
84. Myles R. Allen et al., "A Solution to the Misrepresentations of CO 2-Equivalent Emissions of Short-Lived Climate Pollutants under Ambitious Mitigation," *Npj Climate and Atmospheric Science* 1, no. 1 (June 4, 2018): 1–8, https://doi.org/10.1038/s41612-018-0026-8; Michelle Cain, "Guest Post: A New Way to Assess 'Global Warming Potential' of Short-Lived Pollutants," *Carbon Brief*, June 7, 2018, https://www.carbonbrief.org/guest-post-a-new-way-to-assess-global-warming-potential-of-short-lived-pollutants.
85. John H. Vandermeer and Ivette Perfecto, "Syndromes of Production in Agriculture: Prospects for Social-Ecological Regime Change," *Ecology and Society* 17, no. 4 (2012).
86. Matthew D. Turner, John G. McPeak, and Augustine Ayantunde, "The Role of Livestock Mobility in the Livelihood Strategies of Rural Peoples in Semi-Arid West Africa," *Human Ecology* 42, no. 2 (2014): 231–47.
87. The phrase is from Andreas Malm's unfortunate pamphlet, *Corona, Climate, Chronic Emergency: War Communism in the Twenty-First Century* (Verso Books, 2020); on which, see Max Ajl, "Andreas Malm's Corona, Climate, Chronic Emergency," *The Brooklyn Rail*, November 10, 2020, https://brooklynrail.org/2020/11/field-notes/Corona-Climate-Chronic-Emergency.
88. Brad Ridoutt, "It Takes 21 Litres of Water to Produce a Small Chocolate Bar. How Water-Wise Is Your Diet?," *The Guardian*, October 6, 2019, www.theguardian.com/commentisfree/2019/oct/07/it-takes-21-litres-of-water-to-produce-a-small-chocolate-bar-how-water-wise-is-your-diet.
89. Anne Mottet et al., "Livestock: On Our Plates or Eating at Our Table? A New Analysis of the Feed/Food Debate," *Global Food Security*, Food Security Governance in Latin America, 14 (September 1, 2017): 1–8, https://doi.org/10.1016/j.gfs.2017.01.001.
90. B. Dumont et al., "Prospects from Agroecology and Industrial Ecology for Animal Production in the 21st Century," *Animals* 7, no. 6 (June 2013): 1028–43, https://doi.org/10.1017/S1751731112002418.
91. "The Utilization of Indigenous Knowledge in Range Management and Forage Plants for Improving Livestock Productivity and Food Security in the Maasai and Barbaig Communities," accessed February 7, 2021, www.fao.org/3/a0182e/A0182E07.htm.

92. Thanks to Alex Heffron for imagery and specific arguments here.
93. Richard Teague and Matt Barnes, "Grazing Management That Regenerates Ecosystem Function and Grazingland Livelihoods," *African Journal of Range & Forage Science* 34, no. 2 (2017): 77–86.
94. Max Paschall, "The Lost Forest Gardens of Europe," Shelterwood Forest Farm, accessed October 21, 2020, www.shelterwoodforestfarm.com/blog/the-lost-forest-gardens-of-europe.
95. Andrés Horrillo, Paula Gaspar, and Miguel Escribano, "Organic Farming as a Strategy to Reduce Carbon Footprint in Dehesa Agroecosystems: A Case Study Comparing Different Livestock Products," *Animals* 10, no. 1 (2020): 162.
96. Paige L. Stanley et al., "Impacts of Soil Carbon Sequestration on Life Cycle Greenhouse Gas Emissions in Midwestern USA Beef Finishing Systems," *Agricultural Systems* 162 (May 1, 2018): 249–58, https://doi.org/10.1016/j.agsy.2018.02.003.
97. Tong Wang et al., "GHG Mitigation Potential of Different Grazing Strategies in the United States Southern Great Plains," *Sustainability* 7, no. 10 (October 2015): 13500–521, https://doi.org/10.3390/su71013500.
98. Mimi Hillenbrand et al., "Impacts of Holistic Planned Grazing with Bison Compared to Continuous Grazing with Cattle in South Dakota Shortgrass Prairie," *Agriculture, Ecosystems & Environment* 279 (July 1, 2019): 156–68, https://doi.org/10.1016/j.agee.2019.02.005.
99. Yuting Zhou et al., "Climate Effects on Tallgrass Prairie Responses to Continuous and Rotational Grazing," *Agronomy* 9, no. 5 (May 2019): 219, https://doi.org/10.3390/agronomy9050219.
100. Kat Kerlin, "Grasslands More Reliable Carbon Sink than Trees," *Science and Climate* (blog), July 9, 2018, https://climatechange.ucdavis.edu/news/grasslands-more-reliable-carbon-sink-than-trees/.
101. Hannah Gosnell, Kerry Grimm, and Bruce E. Goldstein, "A Half Century of Holistic Management: What Does the Evidence Reveal?," *Agriculture and Human Values*, January 23, 2020, https://doi.org/10.1007/s10460-020-10016-w.
102. Enrique Murgueitio et al., "Native Trees and Shrubs for the Productive Rehabilitation of Tropical Cattle Ranching Lands," *Forest Ecology and Management*, The Ecology and Ecosystem Services of Native Trees: Implications for Reforestation and Land Restoration in Mesoamerica, 261, no. 10 (May 15, 2011): 1654–63, https://doi.org/10.1016/j.foreco.2010.09.027; Lisa Palmer, "In the Pastures of Colombia, Cows, Crops and Timber Coexist," Yale E360, accessed February 7, 2021, https://e360.yale.edu/features/in_the_pastures_of_colombia_cows_crops_and_timber_coexist.
103. Christian Schader et al., "Impacts of Feeding Less Food-Competing Feedstuffs to Livestock on Global Food System Sustainability," *Journal of The Royal Society Interface* 12, no. 113 (December 6, 2015): 20150891, https://doi.org/10.1098/rsif.2015.0891.

104. Diana Rodgers and Robb Wolf, *Sacred Cow: The Case for (Better) Meat: Why Well-Raised Meat Is Good for You and Good for the Planet*, Illustrated edition (BenBella Books, 2020).
105. Benjamin E. Graeub et al., "The State of Family Farms in the World," *World Development* 87 (November 1, 2016): 1–15, https://doi.org/10.1016/j.worlddev.2015.05.012; Catherine Badgley et al., "Organic Agriculture and the Global Food Supply," *Renewable Agriculture and Food Systems* 22, no. 2 (June 2007): 86–108, https://doi.org/10.1017/S1742170507001640.
106. "La Via Campesina Position Paper: Small Scale Sustainable Farmers Are Cooling Down The Earth – Via Campesina," Via Campesina English, March 25, 2010, https://viacampesina.org/en/la-via-campesina-position-paper-small-scale-sustenable-farmers-are-cooling-down-the-earth/; "Food Sovereignty: Five Steps to Cool the Planet and Feed Its People," accessed October 28, 2020, www.grain.org/article/entries/5102-food-sovereignty-five-steps-to-cool-the-planet-and-feed-its-people.
107. Lauren C. Ponisio et al., "Diversification Practices Reduce Organic to Conventional Yield Gap," *Proceedings of the Royal Society B: Biological Sciences* 282, no. 1799 (January 22, 2015): 20141396, https://doi.org/10.1098/rspb.2014.1396.
108. Mario A. Gonzalez-Corzo, "The Evolution of Agricultural Production and Yields in Post-Reform Cuba," *Economics Bulletin* 39, no. 2 (2019): 1586–1601.
109. Peter Michael Rosset et al., "The Campesino-to-Campesino Agroecology Movement of ANAP in Cuba: Social Process Methodology in the Construction of Sustainable Peasant Agriculture and Food Sovereignty," *The Journal of Peasant Studies* 38, no. 1 (2011): 161–91.
110. Miguel A. Altieri, "Applying Agroecology to Enhance the Productivity of Peasant Farming Systems in Latin America," *Environment, Development and Sustainability* 1, no. 3–4 (1999): 197–217.
111. Marc Barzman and Luther Das, "Ecologising Rice-Based Systems in Bangladesh," *ILEIA Newsletter* 16, no. 4 (2000): 16–17; Marc Barzman and Sylvie Desilles, *Diversifying Rice-Based Farming Systems and Empowering Farmers in Bangladesh Using the Farmer Field-School Approach* (Earthscan, 2013).
112. Roland Bunch, "More Productivity with Fewer External Inputs: Central American Case Studies of Agroecological Development and Their Broader Implications," *Environment, Development and Sustainability* 1, no. 3–4 (1999): 219–33.
113. Zareen Pervez Bharucha, Sol Bermejo Mitjans, and Jules Pretty, "Towards Redesign at Scale through Zero Budget Natural Farming in Andhra Pradesh, India," *International Journal of Agricultural Sustainability* 18, no. 1 (January 2, 2020): 12, https://doi.org/10.1080/14735903.2019.1694465; Ashlesha Khadse et al., "Taking Agroecology to Scale: The Zero Budget Natural Farming Peasant Movement in Karnataka, India," *The Journal of Peasant Studies* 45, no. 1 (January 2, 2018): 192–219, https://doi.org/10.1080/03066150.2016.1276450.

114. Saddam Hossen Majumder, Prodyut Bijoy Gogoi, and Nivedita Deka, "System of Rice Intensification (SRI): An Innovative and Remunerative Method of Rice Cultivation in Tripura, India," *Indian Journal of Agricultural Research* 53, no. 4 (2019): 504–7.
115. Krishna Chaitanya Anantha et al., "Carbon Dynamics, Potential and Cost of Carbon Sequestration in Double Rice Cropping System in Semi Arid Southern India," *Journal of Soil Science and Plant Nutrition* 18, no. 2 (June 2018): 418–34, https://doi.org/10.4067/S0718-95162018005001302.
116. Poornima Varma, "Adoption of System of Rice Intensification under Information Constraints: An Analysis for India," *The Journal of Development Studies* 54, no. 10 (2018): 1838–57.
117. Zeyaur Khan et al., "Push-Pull Technology: A Conservation Agriculture Approach for Integrated Management of Insect Pests, Weeds and Soil Health in Africa: UK Government's Foresight Food and Farming Futures Project," *International Journal of Agricultural Sustainability* 9, no. 1 (2011): 162–70.
118. Harriet Friedmann, "Family Wheat Farms and Third World Diets: A Paradoxical Relationship between Unwaged and Waged Labor," in *Work Without Wages*, ed. Jane L. Collins and Martha Giminez (SUNY Press, 1990), 193–213; Harriet Friedmann and Philip McMichael, "Agriculture and the State System: The Rise and Decline of National Agricultures, 1870 to the Present," *Sociologia Ruralis* 29, no. 2 (1989): 93–117.
119. Chris Smaje, *A Small Farm Future: Making the Case for a Society Built Around Local Economies, Self-Provisioning, Agricultural Diversity and a Shared Earth* (Chelsea Green Publishing, 2020), 160–62.
120. "Agricultural Worker Demographics," NATIONAL CENTER FOR FARMWORKER HEALTH, accessed March 5, 2021, http://www.ncfh.org/agricultural-worker-demographics.html.
121. Kirkpatrick Sale, *Dwellers in the Land: The Bioregional Vision* (University of Georgia Press, 2000).
122. Thomas F. Döring et al., "Evolutionary Plant Breeding in Cereals – into a New Era," *Sustainability* 3, no. 10 (2011): 1961.
123. Monica M. White, *Freedom Farmers: Agricultural Resistance and the Black Freedom Movement* (University of North Carolina Press, 2018); Edward Onaci, *Free the Land: The Republic of New Afrika and the Pursuit of a Black Nation-State* (University of North Carolina Press, 2020), https://muse.jhu.edu/book/74502.
124. Albie Miles, "If We Get Food Right, We Get Everything Right," *Honolulu Civil Beat*, April 11, 2020, www.civilbeat.org/2020/04/if-we-get-food-right-we-get-everything-right/.
125. "Uprooting Racism, Seeding Sovereignty – Schumacher Center for New Economics," accessed February 13, 2020, https://centerforneweconomics.org/publications/uprooting-racism-seeding-sovereignty/.
126. Elizabeth Hoover, "Native Food Systems Impacted by COVID," *Agriculture and Human Values*, accessed October 28, 2020, www.academia.edu/43022886/Native_food_systems_impacted_by_COVID.

127. Devon G. Peña and M. Foucault, "Farmers Feeding Families: Agroecology in South Central Los Angeles," in *Lecture Presented to the Environmental Science, Policy and Management Colloquium*, 2005.
128. Jim Goodman, "We Have Always Had the Solution to Crises Like COVID-19," National Family Farm Coalition, May 7, 2020, https://nffc.net/we-have-always-had-the-solution-crises-covid19/.

CHAPTER 7

1. Vladimir I. Lenin, "Preliminary Draft Theses on the National and Colonial Questions," *Collected Works* 31 (1920): 144–51.
2. Sam Moyo, Praveen Jha, and Paris Yeros, "The Classical Agrarian Question: Myth, Reality and Relevance Today," *Agrarian South: Journal of Political Economy* 2, no. 1 (2013): 93–119; Amilcar Cabral, *Unity and Struggle: Speeches and Writings of Amilcar Cabral* (Monthly Review Press, 1979).
3. Enrique D. Dussel, *Hacia un Marx desconocido: un comentario de los manuscritos del 61–63* (Siglo XXI, 1988), 312–61.
4. It is striking that one of the serious interventions on this front, Corinna Dengler and Lisa Marie Seebacher, "What About the Global South? Towards a Feminist Decolonial Degrowth Approach," *Ecological Economics* 157 (March 1, 2019): 246–52, https://doi.org/10.1016/j.ecolecon.2018.11.019 is silent on climate debt and ecological debt, does not use the word imperialism, and states,"Colonialism and Northern expansionary politics set the basis for today's hierarchically structured global system, in which territorial imperialism has largely been replaced with policies of economic re-structuring under the guise of 'sustainable development.'" This is not a reality-based account of US/EU foreign policy.
5. Ulrich Brand and Markus Wissen, "Crisis and Continuity of Capitalist Society–Nature Relationships: The Imperial Mode of Living and the Limits to Environmental Governance," *Review of International Political Economy* 20, no. 4 (2013): 687–711.
6. M. T. Huber, "Ecological Politics for the Working Class," *Catalyst: A Journal of Theory and Strategy* 3, no. 1 (2019).
7. Collective, "Just Transition" (Transnational Institute, February 2020).
8. UNCTAD, *Financing a Global Green New Deal*, Trade and Development Report (United Nations, 2019).
9. Keston Perry, "Financing a Global Green New Deal: Between Techno-Optimist Renewable Energy Futures and Taming Financialization for a New 'Civilizing' Multilateralism," *Development and Change*, Forthcoming, 9.
10. Glen Sean Coulthard, *Red Skin, White Masks: Rejecting the Colonial Politics of Recognition* (University of Minnesota Press, 2014); Ali Kadri, *Imperialism with Reference to Syria* (Springer, 2019).
11. Archana Prasad, "Ecological Crisis, Global Capital and the Reinvention of Nature," in *Rethinking the Social Sciences with Sam Moyo*, ed. Praveen Jha, Paris Yeros, and Walter Chambati (New Delhi: Tulika Books, 2020), 180–97.

12. Samir Amin, *Accumulation on a World Scale: A Critique of the Theory of Underdevelopment* (Monthly Review Press, 1974).
13. Chris Gilbert, "To Recover Strategic Thought and Political Practice," *MR Online*, September 29, 2015, https://mronline.org/2015/09/29/gilbert290915-html/; Sam Moyo and Paris Yeros, "The Fall and Rise of the National Question," in *Reclaiming the Nation: The Return of the National Question in Africa, Asia and Latin America*, ed. Sam Moyo and Paris Yeros (Pluto Press, 2011), 3–28, https://doi.org/10.2307/j.ctt183hotp.4; Álvaro García Linera, "El Evismo: Lo Nacional-Popular En Acción," *Osal* 7, no. 19 (2006).a, 2011.
14. "People's Agreement of Cochabamba"; "Rights of Mother Earth."
15. Andrew Curley and Majerle Lister, "Already Existing Dystopias: Tribal Sovereignty, Extraction, and Decolonizing the Anthropocene," in *Handbook on the Changing Geographies of the State*, n.d., 251.
16. Sit Tsui et al., "The Development Trap of Financial Capitalism: China's Peasant Path Compared," *Agrarian South: Journal of Political Economy* 2, no. 3 (2013): 247–68.
17. George Manuel, *The Fourth World: An Indian Reality* (University of Minnesota Press, 1974).
18. Matthew Stilwell, *Climate Debt – A Primer* (Third World Network, 2009); Republic of Bolivia, "Commitments for Annex I Parties under Paragraph 1(b)(i) of the Bali Action Plan: Evaluating Developed Countries' Historical Climate Debt to Developing Countries Submission by the Republic of Bolivia to the AWG-LCA," n.d., https://unfccc.int/files/kyoto_protocol/application/pdf/bolivia250409.pdf.
19. United Nations Framework Convention on Climate Change Secretariat, *United Nations Framework Convention on Climate Change* (UNFCCC, 1992).
20. Secretariat.
21. "Final Conclusions Working Group 8: Climate Debt," *World People's Conference on Climate Change and the Rights of Mother Earth* (blog), April 30, 2010, https://pwccc.wordpress.com/2010/04/30/final-conclusions-working-group-n%c2%ba-8-climate-debt/.
22. "Final Conclusions Working Group 8."
23. Republic of Bolivia, "Commitments for Annex I Parties under Paragraph 1(b)(i) of the Bali Action Plan: Evaluating Developed Countries' Historical Climate Debt to Developing Countries Submission by the Republic of Bolivia to the AWG-LCA," n.d., https://unfccc.int/files/kyoto_protocol/application/pdf/bolivia250409.pdf.
24. Rikard Warlenius, "Calculating Climate Debt. A Proposal," 2012, 19–21, www.academia.edu/9167899/Calculating_Climate_Debt_A_Proposal.
25. IPCC, "Chapter 2 – Global Warming of 1.5°C," 80–81.
26. "Submission by the Plurinational State of Bolivia," *World People's Conference on Climate Change and the Rights of Mother Earth* (blog), June 1, 2010, https://pwccc.wordpress.com/2010/06/01/submission-by-the-plurinational-state-of-bolivia-2/.

27. Rikard Warlenius, "Decolonizing the Atmosphere: The Climate Justice Movement on Climate Debt," *The Journal of Environment & Development* 27, no. 2 (2018).
28. Niall McCarthy, "Report: The U.S. Military Emits More CO2 Than Many Industrialized Nations [Infographic]," Forbes, June 13, 2019, https://www.forbes.com/sites/niallmccarthy/2019/06/13/report-the-u-s-military-emits-more-co2-than-many-industrialized-nations-infographic/.
29. Curley and Lister, "Already Existing Dystopias: Tribal Sovereignty, Extraction, and Decolonizing the Anthropocene," 256.
30. Cited in "Bernie Sanders' Climate Plan: Excellent On Electrification, But Concerningly Authoritarian & Populist – #Election2020," *CleanTechnica*, September 28, 2019, https://cleantechnica.com/2019/09/28/bernie-sanders-climate-plan-excellent-on-electrification-but-concerningly-authoritarian-populist-election2020/.
31. Max Ajl, "Report Card on Bernie Sanders' Green New Deal," *Uneven Earth* (blog), August 27, 2019, http://unevenearth.org/2019/08/report-card-on-bernie-sanders-green-new-deal/.
32. "Plan for Climate Change and Environmental Justice | Joe Biden," Joe Biden for President: Official Campaign Website, accessed November 16, 2020, https://joebiden.com/climate-plan/.
33. Green Party of the United States, "The Green New Deal," Green Party of the United States, January 21, 2019, https://gpus.org/organizing-tools/the-green-new-deal/.
34. The Red Nation, "The Red Deal, Part Three: Heal Our Planet," April 27, 2020,9,12,https://therednation.org/the-red-nation-launches-part-three-heal-our-planet-of-the-red-deal/.
35. Cabral, *Unity and Struggle.*
36. Utsa Patnaik, "Revisiting the 'Drain', or Transfers from India to Britain in the Context of Global Diffusion of Capitalism," in *Agrarian and Other Histories: Essays for Binay Bhushan Chaudhuri*, ed. Shubhra Chakrabarti and Utsa Patnaik (Tulika Books, 2017), 277–318; Utsa Patnaik, "Profit Inflation, Keynes and the Holocaust in Bengal, 1943–44," *Economic & Political Weekly* 53, no. 42 (2018): 33; Alec Gordon, "A Last Word: Amendments and Corrections to Indonesia's Colonial Surplus 1880–1939," *Journal of Contemporary Asia* 48, no. 3 (May 27, 2018): 508–18, https://doi.org/10.1080/00472336.2018.1433865; Mike Davis, *Late Victorian Holocausts: El Nino Famines and the Making of the Third World* (Verso Books, 2002).
37. Ruy Mauro Marini, "Subdesarrollo y Revolución" (Siglo Veintiuno Editores Mexico City, 1969).
38. Cabral, *Unity and Struggle*, 130.
39. Max Ajl, "Farmers, Fellaga, and Frenchmen" (PhD, Cornell University, 2019); Sam Moyo, *The Land Question in Zimbabwe* (Sapes Books Harare, 1995); Sam Moyo, "Three Decades of Agrarian Reform in Zimbabwe," *Journal of Peasant Studies* 38, no. 3 (2011): 493–531.

40. Max Ajl, "The Arab Nation, The Chinese Model, and Theories of Self-Reliant Development," in *Non-Nationalist Forms of Nation-Based Radicalism: Nation beyond the State and Developmentalism*, ed. Ilker Corut and Joost Jongerden (Routledge, 2021); Ismail-Sabri Abdallah, "Al-tanmīyya al-mustaqila: muḥāwala litaḥdīd mafhūm mujahal [Independent Development: An Attempt to Define an Unknown Concept]," in *Al-tanmīyya al-mustaqila fi al-waṭan al-'arabī [Independent Development in the Arab Nation]*, ed. Nader Fergany (Center for Arab Unity Studies, 1987), 25–56.
41. Frantz Fanon, *The Wretched of the Earth* (Grove Press, 2007).
42. Max Ajl, "Delinking's Ecological Turn: The Hidden Legacy of Samir Amin," ed. Ushehwedu Kufakurinani, Ingrid Harvold Kvangraven, and Maria Dyveke Styve, *Review of African Political Economy*, no. Samir Amin and Beyond: Development, Dependence and Delinking in the Contemporary World (2021); Samir Amin, *Delinking: Towards a Polycentric World* (Zed Books, 1990).
43. Max Ajl, "The Political Economy of Thermidor in Syria: National and International Dimensions," in *Syria: From National Independence to Proxy War* (Springer, 2019), 209–45; Linda Matar, *The Political Economy of Investment in Syria* (Palgrave Macmillan UK, 2016); Matteo Capasso, "The War and the Economy: The Gradual Destruction of Libya," *Review of African Political Economy* 47, no. 166 (October 1, 2020): 545–67, https://doi.org/10.1080/03056244.2020.1801405.
44. Helen Yaffe, *We Are Cuba!: How a Revolutionary People Have Survived in a Post-Soviet World*, illustrated edition (Yale University Press, 2020).
45. Gowan, *The Global Gamble.*
46. Diana Johnstone, *Fools' Crusade: Yugoslavia, Nato, and Western Delusions*, 1st edition (Monthly Review Press, 2002); Edward S. Herman, David Peterson, and Noam Chomsky, *The Politics of Genocide* (Monthly Review Press, 2010).
47. "Federico Fuentes on Twitter"; Federico Fuentes, "Bolivia: NGOs Wrong on Morales and Amazon," *Green Left* (Green Left, September 6, 2016), Bolivia, www.greenleft.org.au/content/bolivia-ngos-wrong-morales-and-amazon.
48. Archana Prasad, "Ecological Crisis, Global Capital and the Reinvention of Nature," in *Rethinking the Social Sciences with Sam Moyo*, ed. Praveen Jha, Paris Yeros, and Walter Chambati (Tulika Books, 2020), 180–97.
49. Gabor, "The Wall Street Consensus."
50. Batul Suleiman, "al-wahidat al-sha'biyya al-muslaha»... waraqa al-yasar «alrabiha» fi amrika al-latiniyya," *Al-Akhbar*, February 8, 2020, https://al-akhbar.com/World/284364. Thanks to Patrick Higgins for bringing this article to my attention.
51. William Cronon, *Changes in the Land: Indians, Colonists, and the Ecology of New England* (Macmillan, 2011); Carolyn Merchant, *Ecological Revolutions: Nature, Gender, and Science in New England*, First Edition (Chapel Hill: The University of North Carolina Press, 1989).

52. Kyle Whyte, "Indigenous Experience, Environmental Justice and Settler Colonialism," SSRN Scholarly Paper (Social Science Research Network, April 25, 2016), https://doi.org/10.2139/ssrn.2770058.
53. Freedom Mazwi and George T. Mudimu, "Why are Zimbabwe's Land Reforms Being Reversed?," *Economic and Political Weekly* 54 (August 13, 2019); Sam Moyo and Paris Yeros, "Intervention: The Zimbabwe Question and the Two Lefts," *Historical Materialism* 15 (August 31, 2007): 171–204, https://doi.org/10.1163/156920607X225924.
54. Ricardo Jacobs, "An Urban Proletariat with Peasant Characteristics: Land Occupations and Livestock Raising in South Africa," *The Journal of Peasant Studies* 45, no. 5–6 (2018): 884–903.
55. Nick Estes, *Our History Is the Future: Standing Rock Versus the Dakota Access Pipeline, and the Long Tradition of Indigenous Resistance* (Verso Books, 2019); Roxanne Dunbar-Ortiz, *An Indigenous Peoples' History of the United States* (Beacon Press, 2014).
56. Anchorage Declaration, "Indigenous Peoples' Global Summit on Climate Change, Anchorage Alaska, April 24th 2009," *Viewed at www.indigenous portal. com/climate-change/the-anchorage-declaration.html*, 2009.
57. Declaration.
58. International Indigenous Peoples Forum on Climate Change, "Policy Proposals on Climate Change," September 27, 2009, www.forestpeoples.org/sites/default/files/publication/2010/08/iipfccpolicysept09eng.pdf.
59. The Red Nation, "The Red Deal, Part Three: Heal Our Planet."
60. The Red Nation.
61. Richard Schuster et al., "Vertebrate Biodiversity on Indigenous-Managed Lands in Australia, Brazil, and Canada Equals That in Protected Areas," *Environmental Science & Policy* 101 (November 1, 2019): 1–6, https://doi.org/10.1016/j.envsci.2019.07.002.
62. Víctor Toledo, "Indigenous Peoples and Biodiversity," *Encyclopedia of Biodiversity* 3 (January 1, 1999), https://doi.org/10.1016/B978-0-12-384719-5.00299-9.
63. Schuster et al., "Vertebrate Biodiversity on Indigenous-Managed Lands in Australia, Brazil, and Canada Equals That in Protected Areas."
64. Toledo, "Indigenous Peoples and Biodiversity."
65. Monica Evans, "Respect for Indigenous Land Rights Key in Fight against Climate Change," CIFOR Forests News, September 24, 2020, https://forestsnews.cifor.org/67515/respect-for-indigenous-land-rights-key-in-fight-against-climate-change?fnl=en.
66. "Saving Caribou and Preserving Food Traditions Among Canada's First Nations," *Civil Eats*, October 29, 2020, https://civileats.com/2020/10/29/saving-caribou-and-preserving-food-traditions-among-canadas-first-nations/.
67. Tony Marks-Block, "Indigenous Solutions to California's Capitalist Conflagrations," *MR Online* (blog), October 23, 2020, https://mronline.org/2020/10/23/indigenous-solutions-to-californias-capitalist-conflagrations/.

CONCLUSION

1. Philip McMichael, *Development and Social Change: A Global Perspective* (SAGE Publications, 2011); Gabriel Kolko, *Confronting the Third World: United States Foreign Policy, 1945-1980* (Pantheon Books, 1986); Vincent Bevins, *The Jakarta Method: Washington's Anticommunist Crusade and the Mass Murder Program That Shaped Our World* (New York: PublicAffairs, 2020).
2. Ali Kadri, *China's Path to Development: Against Neoliberalism* (Springer Singapore, 2021), https://doi.org/10.1007/978-981-15-9551-6.

Index